T0348835

REFLECTION SEISMOLOGY: THEORY, DATA PROCESSING AND INTERPRETATION

REFLECTION SEISMOLOGY: THEORY, DATA PROCESSING AND INTERPRETATION

WENCAI YANG

AMSTERDAM • BOSTON • HEIDELBERG • LONDON
NEW YORK • OXFORD • PARIS • SAN DIEGO
SAN FRANCISCO • SINGAPORE • SYDNEY • TOKYO

Elsevier

225, Wyman Street, Waltham, MA 02451, USA

The Boulevard, Langford Lane, Kidlington, Oxford OX5 1GB, UK

Radarweg 29, PO Box 211, 1000 AE Amsterdam, The Netherlands

Notice

No responsibility is assumed by the publisher for any injury and/or damage to persons or property as a matter of products liability, negligence or otherwise, or from any use or operation of any methods, products, instructions or ideas contained in the material herein. Because of rapid advances in the medical sciences, in particular, independent verification of diagnoses and drug dosages should be made

Library of Congress Cataloging-in-Publication Data

Wencai Yang, 1942-

   Reflection seismology : theory, data processing, and interpretation / Wencai Yang.

      pages cm

   Includes bibliographical references.

   ISBN 978-0-12-409538-0 (hardback)

1. Seismic reflection method. I. Title.

   TN269.84.Y26 2013

   551.22–dc23

2013029456

British Library Cataloguing in Publication Data

A catalogue record for this book is available from the British Library

ISBN: 978-0-12-409538-0

For information on all **Elsevier** publications
visit our web site at store.elsevier.com

**Working together
to grow libraries in
developing countries**

www.elsevier.com • www.bookaid.org

Contents

Preface

Reflection seismology belongs to a small branch of solid Earth physics. Does it have its own theoretical systems? This is the question that I keep considering. If one copies word by word the formula from elastic or continuum mechanics to form the theory of reflection seismology, one looks upon reflection seismology as a branch of applied technology, but not as a branch of applied science. Even today, many geoscientists still regard reflection seismology as engineering technology that surveys mineral exploration. As reflection seismologists, we cannot blame their ignorance, but can only blame ourselves for not having built the theoretical system perfectly and completely.

It is true that reflection seismology was an oil/gas exploration technique at the beginning. An oil field was found according to the seismic reflection data in the continent of Oklahoma in the USA in 1926, which proves that seismic reflection is really a good practical technique having enormous economic worth. After the 1930s, based on decades of effort, especially after the invention of computers, the theory of reflection seismology has been developed rapidly. Some monographs on reflection seismology have been published in the 1970s. However, because the renewal speed of seismic exploration methods has been too fast, these monographs cannot include enough newly developed content on the theory of reflection seismology. On the other hand, textbooks used in universities mostly quote elastic mechanics formulas, and have not absorbed enough newly developed theories on reflection seismology. From 1983 to 2013, I have written a few monographs on geophysical inversion theory and methods that have been my main research area. As the foundation of performing seismic inversion is built on solving the forward problems in reflection seismology, I have devoted my time for summarizing the theory of reflection seismology, merging forward and inverse problems into a textbook, and bringing out the best in each other.

This book is designed as a textbook for professional graduate students majoring in applied geophysics. It is impossible to cover all developments in reflection seismology theories. I have outlined just the equation systems of reflection seismology based on the theory of continuum mechanics. The equation family constructs the theoretical framework of reflection seismology. I do not want to emphasize the abstract beauty of these equations, but in this book, I try to put up systematically a skeleton that is formed with equations and mathematical formulae, showing the spirit and values of classical physics. Equations and mathematical formulae are the essential language for communication between human beings and computers, applicable to all. One can be an excellent geologist who may not be good at mastering this kind of language, but one cannot be an excellent seismologist who is not good at mastering this kind of language.

The corresponding teaching of how to use this textbook takes about 50 h. As teaching materials, the authors must consider the following three points and guard against three misleading factors, namely first, write teaching materials as research reports, discuss a lot of details, and mislead students to ignore the backbone. Second, write the textbook as a collection of practical techniques that will be updated soon, and do not probe into their theoretical bases deeply. Students tend to listen excitedly, think that a lot of useful knowledge would be obtained, but ignore the academic ability and marrow. Third, write the teaching materials as an encyclopedia that includes every aspect of the subject, but not fully systematically. Books become very voluminous when they provide too much of knowledge for students, but become unfavorable for training graduate students who have a systematic thinking ability. To avoid misleading the readers, this book adheres to the main idea, and does discuss the main branches in detail; it only explains the theory and not already existing market-based technology. I am not afraid of writing little, but only afraid of writing too much. The backbone must explain clearly, the minor branches may only be clicked. It not only lets students assimilate information but also confers the ability to combine bits of knowledge to solve practical problems in application.

This book is divided into 7 chapters as follows: The first chapter reviews the basic wave theory of classical physics, and the second summarizes elastic wave equations that build up the foundation of reflection seismology. Particle dynamics used to be improperly applied to explain vibration and wave propagation in the past, and I have tried my best to correct improper concepts by using continuum mechanics. Chapter three summarizes seismic wave propagation in the solid Earth, especially some special characteristics different from elastic waves in common solid media. Chapter four deals with wave equation changes along with seismic data processing, together with changes on the relevant boundary conditions. Chapter five discusses the integral solutions of wave propagation problems and Green's function methods. Chapter six discusses the decomposition and continuation of seismic wave fields, with emphasis on the properties of wave equations with variable coefficients and the operator expansion method for the problems under study. The last chapter briefly introduces inverse problems involved in seismic exploration and typical solving numerical methods.

I would like to thank professors Fu Chengyi and W. Telford for guiding me on solid Earth geophysics. I also appreciate the help of Drs. Yang Wuyang, Li Lin, Xie Chunhui, Wang Enli, Wang Wanli, and Zhu Xiaoshan in the translation of this book to English. The author is much indebted to Mohanapriyan Rajendran for improving the English presentation of this book and grateful to all colleagues and readers who inform errors left in this book.

Wencai Yang
National Lab on Tectonics and Dynamics,
Institute of Geology, CAGS.PRC

CHAPTER

1

Introduction to the Wave Theory

OUTLINE

Physics is the study of the motion of matter. It originates from Newtonian mechanics theory. In classical mechanics, which is established by Newton in the seventeenth century, particle dynamics was first employed to describe the motion of macroscopic objects. However, we must deal with continuously constructed media such as the earth; the application of particle dynamics has some limitations. Thus, scientists developed continuum mechanics in the early twentieth century, including fluid mechanics and solid mechanics. Quantum mechanics has been developed in the same period to describe the motion of microscopic particles. The theory in this book is based on continuum mechanics and the methods of mathematical physics.

Reflection Seismology
http://dx.doi.org/10.1016/B978-0-12-409538-0.00001-4

1

The motion of macroscopic objects in general can be divided into three types: The first is displacement, such as linear movement, rotation, flight, and flow. The second is vibration and wave motion, such as periodic motion, water wave, acoustic wave, and seismic wave. The third is chaotic movement, such as intermittent motion, turbulence, and nonlinear wave motion. This book only discusses classical vibration and wave theory. As a kind of physical movement, vibration can be described using an initial value problem of ordinary differential equations for a closed system, in which energy and information do not get exchanged between the system and the outside world. For an open system, in which energy and information get exchanged between the system and the outside world, the initial and boundary value problems of partial differential equations must be applied. This chapter discusses the basic wave theory, focusing on the acoustic wave equation and related wave behaviors.

To make mathematical formulas clear, we use bold English letters for vectors or matrices, and normal Greek letters for scalars in this book.

1.1. WAVE MOTION IN CONTINUOUS MEDIA

Wave motion refers to the propagation of vibration in continuous media, which can be described by using the following formulations:

Wave motion = vibration + propagation (in continuous media);
Vibration = periodic motion that an object moves around its equilibrium point;
Propagation = interaction between the vibration and adjoining mass grains and diffusion of the vibration energy.

The above concepts are limited to the situation that the vibration equilibrium point of mass grains is fixed and is stationary during wave propagation; they are accurate for the elastic waves propagating in solid media. Generalized waves involve situations in which the vibration equilibrium point is not fixed and movable. For example, water waves are a kind of generalized waves with moving equilibrium points. We do not discuss generalized waves in this book.

Three assumptions are usually accepted in the study of continuum mechanics and are as follows (Fung, 1977; Spencer, 1980; Du Xun, 1985):

1. Mass motion follows the law of conservation of mass, that is the first derivative of mass M with respect to time t is zero:

$$\frac{\mathrm{d}M}{\mathrm{d}t} = 0 \tag{1.1}$$

2. The equilibrium point of mass–grain vibration in a wave field is fixed and stable.

3. Vibration occurs in continuous media, and the interaction between adjacent mass grains follows the continuity equation, which defines the continuous media.

What kind of media can be defined as continuous media? In other words, what kind of mechanical laws are followed by continuous media?

Definition: for a small deformation, continuous media are defined by the continuity equation as follows:

$$\rho = \rho_0(1 - div\ \mathbf{u}) \tag{1.2}$$

In this equation, ρ denotes the density of a mass grain in the media with respect to time t; ρ_0 denotes the density of the mass grain with respect to an initial time t_0; $\mathbf{u}$ denotes the displacement vector of the mass grain; $div\ \mathbf{u}$ indicates the divergence of the displacement vector. Equation (1.2) could be derived from Eqn (1.1), which describes the conservation of mass. The product of the grain volume and density is called the mass element.

Set V as the volume of continuous media and the mass in Eqn (1.1) can be expressed as follows:

$$M = \int_V \rho dV$$

In Cartesian coordinates, according to the law of the conservation of mass, mass can be expressed as follows:

$$M = \int_V \rho\ dV = \int_V \rho_0 dV \tag{1.3}$$

where ρ_0 represents the density at the initial time t_0. Because $dv = dx\ dy\ dz$ and $dv_o = dx_o\ dy_o\ dz_o$, denoting a determinant as

$$J = \begin{vmatrix} \dfrac{\partial x}{\partial x_0} & \dfrac{\partial x}{\partial y_0} & \dfrac{\partial x}{\partial z_0} \\[2mm] \dfrac{\partial y}{\partial x_0} & \dfrac{\partial y}{\partial y_0} & \dfrac{\partial y}{\partial z_0} \\[2mm] \dfrac{\partial z}{\partial x_0} & \dfrac{\partial z}{\partial y_0} & \dfrac{\partial z}{\partial z_0} \end{vmatrix} \tag{1.4}$$

We have

$$dV = J\ dV_0$$

Substituting Eqn (1.4) into Eqn (1.3) yields

$$\int_{V_0} \rho J\ dV_o = \int_{V_0} \rho_o dV_o \tag{1.5}$$

or

$$\int_{Vo} (\rho J - \rho_o)\mathrm{d}V_o = 0 \tag{1.6}$$

Hence, the integral kernel should be equal to zero as

$$\rho_0 = \rho J \tag{1.7}$$

Suppose a small deformation is being generated during the wave motion. Then,

$$\left|\frac{\partial u_x}{\partial y}\right| \ll 1 \quad \text{and} \quad \left|\frac{\partial u_x}{\partial z}\right| \ll 1, \text{ etc.}$$

If one ignores the second-order terms, Eqn (1.3) can be transformed as

$$J = 1 + \frac{\partial u_x}{\partial x} + \frac{\partial u_y}{\partial y} + \frac{\partial u_z}{\partial z} = 1 + div\ \mathbf{u} \tag{1.8}$$

If we substitute Eqn (1.8) into Eqn (1.7), the continuous Eqn (1.2) can be obtained.

It is true that wave motion involves only small deformations, but explosions involve large deformations. In seismic explorations, explosions are used as the vibration sources to produce seismic waves. Both vibration and movement of mass grains occur around the shots, where one has to use the theory of explosion that we do not discuss in this book. We will explain the movement in the far field area, where the equilibrium points are fixed and stable and small deformations are accepted.

The continuity Eqn (1.2) implies that the density of the mass elements may vary during wave motion and that the variation amplitude is proportional to the divergence of the displacement vector.

The movements of the mass grains in nature can be classified into several kinds, and vibration and wave motion are two of them. There are many physical branches that describe the movements of the mass grains, including classical mechanics, continuum mechanics, and nonlinear dynamics. Classical mechanics is the study of mass motion in free space, usually without additional constraints, and is helpful for the study of gas motion in a vacuum and the Brownian motion of molecules. Mass movement follows Newton's laws of motion and universal gravitation in an enclosed dynamic system and can be expressed by some ordinary differential equations with initial values. Continuum mechanics studies constrained motion in continuum space, it is based on Newton's equation of motion and specific constitutive equation, and it can be described by partial differential equations with initial and boundary values. The constitutive equation in elastic mechanics is called the generalized Hooke's law. Continuity Eqn (1.2) is the starting point of continuum mechanics, which used to be divided into statics and dynamics. Wave motion is the result of some forces, and belongs to dynamics.

1.2. VIBRATION

Vibration can be described as the motion of a mass constraint to an equilibrium point with a limited distance. No matter how the mass oscillates, it will always return to the equilibrium point, due to the direction of the force acting on the oscillator always pointing to the equilibrium point. Only when the direction of force (i.e. elastic force in a solid) is opposite that of the movement, the vibration can be always around the equilibrium point. Therefore, vibration is a kind of mass motion whose working force and displacement are in reverse directions.

In the case of one-dimensional (1D) motion, we denote the elastic force as F, and the displacement of movement is indicated as u; then, Hooke's law is

$$F = -ku \tag{1.9}$$

where k indicates the elastic coefficient, the negative sign indicates that the direction of the force is opposite to the displacement. Denoting the mass of the oscillator as m, we can apply the equation using Newton's second law as (Budak et al., 1964)

$$m\frac{d^2u}{dt^2} = -ku = F \tag{1.10}$$

Setting the circular frequency as ω, we substitute $\omega^2 = \frac{k}{m}$ into Eqn (1.10) and obtain

$$\frac{d^2u}{dt^2} + \omega^2 u = 0 \tag{1.11}$$

The above equation is the vibration equation without damping, and its general solution is

$$u = A\cos(\omega t + \phi) \tag{1.12}$$

where A indicates the amplitude and ϕ indicates the vibration phase.

Any movement will be affected by resistance, and a vibration with resistance is called the damping vibration. Damping comes from the constraints of surrounding media and creates a resistance f_r, which is proportional to the speed of the oscillation when the speed is not too high and their directions are opposite. If we denote γ as the proportional coefficient, then

$$f_r = -\gamma\frac{du}{dt} \tag{1.13}$$

Therefore, Eqn (1.10) can be rewritten as follows:

$$m\frac{\mathrm{d}^2 u}{\mathrm{d}t^2} = -ku - \gamma\frac{\mathrm{d}u}{\mathrm{d}t} = -F - \gamma\frac{\mathrm{d}u}{\mathrm{d}t} \tag{1.14}$$

where $\omega_0^2 = \frac{k}{m}$ indicates the square of the angular frequency of the vibration, and $\beta = \frac{\gamma}{2m}$ is the damping coefficient of the vibration. The damping vibration equation can be written as

$$\frac{\mathrm{d}^2 u}{\mathrm{d}t^2} + 2\beta\frac{\mathrm{d}u}{\mathrm{d}t} + \omega_0^2 u = 0 \tag{1.15}$$

Equation (1.15) is a second-order ordinary differential equation. If $\beta < \omega_0$, its general solution is

$$u = A_0 e^{-\beta t}\cos(\omega t + \phi_0) \tag{1.16}$$

where A_0 and ϕ_0 are the integral constants to be chosen according to the initial condition of vibration. The angular frequency $\omega = \sqrt{\omega_0^2 - \beta^2}$ is related to the damping coefficient.

The vibration of rock grains in reflection seismology mostly belongs to damping oscillations. Equation (1.16) shows that there are two outstanding features in real oscillations. The first is that the magnitude of the amplitude decreases according to the exponential law with respect to increasing time. The second is that the angular frequency ω changes during the damping oscillations, that is the dispersion. Compared with the intrinsic frequency ω_0, the bigger the damping coefficient β is, the lower the frequency ω turns out to be.

1.3. PROPAGATION AND DIFFUSION

Vibration refers to the motion around an equilibrium point, while propagation refers the motion with energy spreading from a source in space. The existence of any object affects the surrounding mass, no matter whether these objects connect with each other or not. A physicist sitting outside a house would feel at least four kinds of forces acting toward him: the gravity from the earth, the pressure from the air, the thermal force from the sun, and the electric or magnetic forces from an antenna or transmission wires. The universe is made of all kinds of materials, and the effect between materials sustains the movements of the universe. Thus, there exists forces everywhere, and the space filled with forces is physically called the field. Forces change the movement of objects, so we can predict the magnitude and direction of a force by measuring the movement of objects (i.e. distance and velocity). Energy exists everywhere in the universe, and it can be embodied in matter, antimatter, or in the field.

How is the field of forces generated? It is generated by the approaching or the entering of some kind of force sources. Explosions and vibrators are the force sources that produce reflective waves, and the vibration of a vibrator propagates through the interaction of molecules in the media.

Propagation is a natural process caused by force sources. Force sources are of two basic types: long-effecting sources and pulse-like sources whose energy can be consumed within just a minute. The long-effecting sources work steadily, making wave propagation more stable. The energy of the second kind can propagate and finally fill the whole space, such as a seismic wave field. On the other hand, force can also be divided into two basic types: body force and surface force. Gravitational and electromagnetic forces belong to body force, and elastic force, which maintains seismic wave motion, and frictional force belongs to surface force. Body forces establish an interaction between substances via the field, but surface forces establish interaction directly via substance contact.

The word "Propagation" is similar to the word "diffusion", and mathematically, these two words are synonymous. In physics, we use "wave propagation" to focus on rather instant processes, use "diffusion" to focus on particle motion, and we use "heat conduction" to focus on rather slow processes. In mathematical physics, we can use the Laplace operator to describe the spatial change of a field in space, which can be used to describe wave propagation, diffusion, and heat conduction. However, they are different with respect to time-varying speed. Diffusion should be described by the first-order derivative with respect to time, resulting in the so-called parabolic equations. Wave propagation may be described by second-order derivatives, resulting in the so-called hyperbolic equation.

We use the gravitational field as an example to look at the energy spreading from a source in space. All substances tend to show their existence to their surroundings. When a car goes into a campus, the buildings, plants, animals, and other facilities would be attracted as the gravitational field caused by the car affects them. Why do these inanimate objects have the same tendency as human beings do? This is a philosophical question. Physical scientists do not answer this question, but only admit the fact and discuss its behavior and related interactions. The spatial diffusion operator of force is described by the divergence in physics (Tiknonov and Samarskii, 1963). Therefore, a quantitative description of the diffusion is

$$divF = div(gradU) = \nabla^2 U \tag{1.17}$$

where F is the gravitational force. U is the potential of the gravity field, and ∇^2 is the Laplace operator (Hutson, 1980), which is equal to

$$\nabla^2 = \frac{\partial^2}{\partial x^2} + \frac{\partial^2}{\partial y^2} + \frac{\partial^2}{\partial z^2}$$

Setting the left-hand side of Eqn (1.17) equal to zero results in the famous Laplace equation.

The Laplace operator contains differentials of the potential with respect to the space coordinates. As stated above, the wave propagation process and diffusion process should follow the same mathematical form. Therefore, we can conclude that there exists a similar term in the wave equation as shown in Eqn (1.17). In essence, the propagation of the wave energy is via spatial diffusion of the equilibrium points of vibration. If a vibration source is located in the sea, a pressure field will be caused by sea water. The spatial change of the pressure field will be described by the divergence in Eqn (1.17). Comparing with the gravitational field, wave propagation must deal with both vibration and displacement with respect to time.

If a substance exists in space, it would announce its existence through the "field" to prevent a collision with other substances. Does it cost its own energy? The answer is yes, although the energy released for the announcement is very little. All substances aging and ancient paintings become old and their canvas loses its elasticity. The fact that the universe expands proves that the universe has been aging, and has also proved the second law of thermodynamics and law of conservation of energy.

In the process of wave propagation, one must consider not only the divergence of the wave field potential generated by force sources but also the motion of vibration. The movement velocity of the vibration equilibrium point is the most important parameter for the description of wave propagation.

1.4. ACOUSTIC WAVE EQUATION

Now we start to consider the most simple equation in wave motion. According to the basic concept of "wave = vibration + propagation", the wave equation must include vibration Eqn (1.11) and propagation operator Eqn (1.17). In Eqn (1.11), the displacement $U(t)$ is a function of time, while in Eqn (1.17), $U(x)$ is a function of spatial coordinates. We have to combine the two functions for obtaining the wave equations.

Let $U(\mathbf{x}, t)$ be a vector function of variables x and t. It has three components denoted as (u, v, w) in 3D homogeneous infinite space. The gradient of the wave function is the strain vector, having its three components as follows:

$$E_x = grad\ \mathrm{u}$$
$$E_y = grad\ \mathrm{v}$$
$$E_z = grad\ \mathrm{w}$$

According to Hooke's law, Stress is proportional to strain:

$$A_x = (\lambda + 2\mu)grad \; \mathbf{u}$$
$$A_y = (\lambda + 2\mu)grad \; \mathbf{v} \qquad (1.18)$$
$$A_z = (\lambda + 2\mu)grad \; \mathbf{w}$$

where λ and μ are Lame's constants. In continuous media, stress divergence is equivalent to body force. In the condition of missing real body force, the equation of motion can be written as follows (see the proof of Eqn (2.13) given later):

$$\rho \frac{\partial^2 \mathbf{u}}{\partial t^2} = div \; \mathbf{A} \qquad (1.19)$$

It is supposed that density and Lame's coefficients are constants. Substituting Eqn (1.18) into Eqn (2.13) yields

$$\nabla^2 u - \frac{1}{c_p^2}\frac{\partial^2 u}{\partial t^2} = 0 \qquad (1.19a)$$

$$\nabla^2 v - \frac{1}{c_p^2}\frac{\partial^2 v}{\partial t^2} = 0 \qquad (1.19b)$$

$$\nabla^2 w - \frac{1}{c_p^2}\frac{\partial^2 w}{\partial t^2} = 0 \qquad (1.19c)$$

where c_p is the p-wave velocity,

$$c_p = \sqrt{\frac{2\lambda + \mu}{\rho}}$$

If one denotes the phase velocity of wave propagation in homogeneous media as c and a body force as $F(x,t)$, then Eqn (1.19a) becomes (Tiknonov and Samarskii, 1963)

$$\frac{\partial^2 u}{\partial x^2} + \frac{\partial^2 u}{\partial y^2} + \frac{\partial^2 u}{\partial z^2} - \frac{1}{c^2}\frac{\partial^2 u}{\partial t^2} = -F \qquad (1.19d)$$

Equation (1.19d) is the scalar acoustic wave equation, a very basic equation commonly used for the description of wave propagation in space. In the next chapter, we will simplify Navier equations to derive this wave equation (Eqn (2.16)) and its vector equation (Eqn (2.17)).

Similarly, if we take the resistance force into account, and add the space propagation operator to damping vibration Eqn (1.14), the result is (Futterman, 1962; Christenson, 1982)

$$\frac{\partial^2 u}{\partial x^2} + \frac{\partial^2 u}{\partial y^2} + \frac{\partial^2 u}{\partial z^2} - \frac{1}{c^2}\frac{\partial^2 u}{\partial t^2} - 2\beta\frac{\partial u}{\partial x} = -F \qquad (1.20a)$$

where β is the damping factor in homogeneous continuous media, which is related to the quality factor Q. The higher the quality factor Q of the media, the lesser is the wave energy absorbed by the media. Therefore, β is inversely proportional to Q. In fact, the wave is propagation of vibrations in space, the wave energy absorbed by the media in the propagation must also relate to both the vibration frequency and the wave velocity. The higher the vibration frequency or velocity, the greater the wave energy absorbed. Therefore, the wave energy absorption coefficient can be defined as

$$\alpha = 2\beta c = \omega/2cQ,$$

and the damping wave Eqn (1.20a) becomes

$$\frac{\partial^2 u}{\partial x^2} + \frac{\partial^2 u}{\partial y^2} + \frac{\partial^2 u}{\partial z^2} - \frac{1}{c^2}\frac{\partial^2 u}{\partial t^2} - \frac{\alpha}{c}\frac{\partial u}{\partial x} = -F \qquad (1.20b)$$

Equations (1.19d) and (1.20b) describe an acoustic wave propagating in a three-dimensional (3D) space. The wave function is a scalar function, which means pressure, elastic energy, etc. together with a source function expressed on the right side of the equation. Assume that the media is homogeneous and that wave velocity c and quality factor Q are both constant. If c and Q are not constant, and it becomes a function of the spatial coordinates, can the acoustic wave Eqns (1.19d) and (1.20b) be established? This question will be discussed in Chapter 3.

Now, we discuss the general solution of the acoustic wave Eqn (1.19d). In addition, we assume that a pulse point source is located at the origin and then the wave field spreads all around the homogeneous media. In the spherically symmetric spherical coordinates (r, t), the acoustic wave equation can be rewritten as

$$\frac{\partial^2 u}{\partial r^2} + \frac{2}{r}\frac{\partial u}{\partial r} - \frac{1}{c^2}\frac{\partial^2 u}{\partial t^2} = \delta(0,0) \qquad (1.21)$$

where

$$u = \frac{1}{r}\{f(r - ct) + g(r + ct)\} \qquad (1.22)$$

If we substitute Eqn (1.22) into Eqn (1.21), we can prove the general solution Eqn (1.22). The point source located at the origin stimulates the wave field $f(r, t)$ whose amplitude decays with $1/r$. The corresponding wave field $g(r, t)$ is the wave field that comes back from infinity. Mathematical solution Eqn (1.22), including $g(r, t)$, indicates that there are echo waves from infinity. If the area studied is limited and the infinity echo is ignored, then we have

$$u^{in} = \frac{1}{r} g(r + ct) = 0 \tag{1.23}$$

This is called the Sommerfeld radiation condition.

If the source is a plane wave but not a point source, we can define the incident direction vector n that is a unit vector. The acoustic wave Eqn (1.21) in spherical coordinates (r, t) becomes

$$\frac{\partial^2 u}{\partial r^2} + \frac{2}{r} \frac{\partial u}{\partial r} - \frac{1}{c^2} \frac{\partial^2 u}{\partial t^2} = 0 \tag{1.24}$$

Its general solution becomes

$$u(r, t) = f(n \cdot r - ct)$$

In Cartesian coordinates, the general solution is

$$u(r, t) = e^{i(k \cdot r - \omega t)} \tag{1.25}$$

where k is a wave vector, satisfying $k \cdot k = \dfrac{\omega^2}{c^2}$.

1.5. ACOUSTIC WAVE EQUATION WITH COMPLEX COEFFICIENTS

Mathematicians like to dabble with their digital language to make every notation complete and simple. Now, we are going dabble in such a way that the acoustic wave equation is unchanged, and the real wave velocity is changed to complex velocity. We will show that the change from the real coefficients to complex coefficients yields the damping acoustic wave equation, which corresponds to wave propagation in viscoelastic media (Christenson, 1982). During discussion of the acoustic equation with complex coefficients, we reveal an important parameter of a substance, that is the quality factor that describes the property of a medium rather than the behavior of wave propagation.

1.5.1. Complex Elastic Modulus and the Complex Wave Velocity

In the real coefficient wave equation, the denominator of the coefficients is the phase velocity c, and it can be expressed as

$$c = \sqrt{\frac{M}{\rho}}; \quad M_p = \lambda + 2\mu; \quad M_s = \mu \tag{1.26}$$

where M is the modulus of elasticity, which can be taken as M_p for the P-wave and as M_s for the S-wave; λ, μ are the Lame constants. In the wave equation with complex coefficients, to define an elastic mass model in which the P-wave and S-wave are not coupled, the complex elastic modulus will be defined as follows:

$$M(i\omega) = \left|M\right|e^{i\delta(\omega)} = M'(i\omega) + iM''(i\omega) \tag{1.27}$$

where δ represents the loss angle and ω represents the circular frequency. Denoting the real and imaginary parts of the elastic modulus as M' and M'', respectively, the quality factor of a medium equals

$$Q(\omega) = \frac{M'(\omega)}{M''(\omega)} = \frac{1}{\tan \delta(\omega)} \tag{1.28}$$

Thus, the Hooke law in the viscoelastic media can be rewritten as

$$A(\omega) = M(i\omega)E(\omega) \tag{1.29}$$

where A is the stress and E is the strain.

To define the 1D complex wave velocity, substituting Eqn (1.29) into Newton's second law yields

$$\rho \frac{\partial^2 u}{\partial t^2} = M(i\omega)\frac{\partial^2 u}{\partial x^2} \tag{1.30}$$

If we compare Eqn (1.30) and Eqn (1.19), we see that the complex wave velocity should be defined as

$$V(i\omega) = \sqrt{\frac{M(i\omega)}{\rho}} = \sqrt{\frac{|M(i\omega)|e^{i\delta(\omega)/2}}{\rho}} \tag{1.31}$$

Thus, Eqn (1.30) can be rewritten as

$$\frac{\partial^2 u}{\partial x^2} - \frac{1}{V^2(i\omega)}\frac{\partial^2 u}{\partial t^2} = 0 \tag{1.32}$$

This equation appears to have the same form as that of the wave equation with real coefficients (Eqn (1.19)), but it has a different meaning. The general solution of Eqn (1.32) is

$$u(x,t) = u_0(\omega)\exp[i\omega t - x/V(i\omega)] \tag{1.33}$$

One has to separate the real part and the imaginary part of the complex wave $u(x,t)$ to find the different meaning. From Eqn (1.31), the complex wave velocity $V(i\omega)$ can be decomposed as

$$\frac{1}{V(i\omega)} = \left[\sqrt{\frac{|M|}{\rho}}\left(\cos\frac{\delta}{2} + i\sin\frac{\delta}{2}\right)\right]^{-1} = \sqrt{\frac{\rho}{|M|}}\left(\cos\frac{\delta}{2} - i\sin\frac{\delta}{2}\right) \tag{1.34}$$

Therefore, the complex slowness can be written as

$$\frac{1}{V(i\omega)} = \frac{1}{c(\omega)} - i\frac{\alpha(\omega)}{\omega} \tag{1.35}$$

The phase velocity can be written as

$$c(\omega) = \sqrt{\frac{|M|}{\rho}} \sec\frac{\delta}{2} \tag{1.36}$$

The absorption coefficient can be rewritten as

$$\alpha = \sqrt{\frac{\rho}{|M|}} \omega \sin\frac{\delta}{2} = \omega \tan\frac{\delta}{2}/c(\omega) \tag{1.36a}$$

The quality factor Q of most rocks is usually >100, and the loss angle becomes very small. Thus, for the wave attenuation coefficient we can write

$$\alpha = \frac{\omega}{2c}\tan\delta = \frac{\omega}{2Qc} \tag{1.37}$$

and

$$Q(\omega) = \frac{\omega}{2\alpha(\omega)c(\omega)} \tag{1.38}$$

From Eqn (1.34), we can see that the real part of the complex slowness is the reciprocal of the phase velocity, and the imaginary part is proportional to the absorption coefficient that is related to the viscoelasticity of media. From Eqn (1.35), we can see that the phase velocity is not only related to the elasticity of the media but is also related to the absorption coefficient of the media. From Eqn (1.37), we can see that the absorption coefficient of the viscoelastic media is related to the quality factor as well as to the phase velocity and the frequency. The above equations indicate that complex wave velocity is a characteristic parameter that can be used to characterize wave propagation in some imperfect elastic media.

1.5.2. Damping Wave Equations in Viscoelastic Media

We are going to derive 3D acoustic equation with complex coefficients and prove that the complex coefficient acoustic equation is equivalent to the damping wave Eqn (1.20). After using Fourier transform with respect to time t, $u(s,y,z,t)$ in Eqn (1.14) turns into $u(x,y,z,\omega)$, so the wave equation in the frequency domain will be written as

$$\frac{\partial^2 u}{\partial x^2} + \frac{\partial^2 u}{\partial y^2} + \frac{\partial^2 u}{\partial z^2} + \frac{\omega^2}{V(i\omega)^2}u = -F \tag{1.39}$$

Substituting Eqn (1.35) into Eqn (1.39) yields

$$\frac{\partial^2 u}{\partial x^2} + \frac{\partial^2 u}{\partial y^2} + \frac{\partial^2 u}{\partial z^2} + \frac{\omega^2 u}{c^2}\left[1 - \frac{1}{4Q^2}\right] - \frac{2i\alpha}{c}\omega u = -F \qquad (1.40)$$

Because

$$\frac{1}{4Q^2} \ll 1$$

Assuming that both the phase velocity and quality factor would not disperse in the frequency band of seismic reflection, Eqn (1.40) in the frequency domain can be rewritten as

$$\frac{\partial^2 u}{\partial x^2} + \frac{\partial^2 u}{\partial y^2} + \frac{\partial^2 u}{\partial z^2} + \frac{\omega^2 u}{c^2} - \frac{2i\alpha}{c}\omega u = -F \qquad (1.41)$$

Finally, the inverse Fourier transform of Eqn (1.41) yields

$$\frac{\partial^2 u}{\partial x^2} + \frac{\partial^2 u}{\partial y^2} + \frac{\partial^2 u}{\partial z^2} - \frac{\omega^2}{c^2}\frac{\partial^2 u}{\partial t^2} - \frac{2\alpha}{c}\frac{\partial u}{\partial t} = -F \qquad (1.20b)$$

This is the damping wave equation Eqn (1.20). The last term on the left side of the equation shows attenuation of a wave field in viscoelastic media. In the equation, the phase velocity c and absorption coefficient α are defined by Eqns (1.36) and (1.36a). Now, we have proved that the complex coefficient acoustic Eqn (1.39) is equivalent to the damping wave Eqn (1.20b).

1.5.3. Viscoelastic Models

We have discussed the improvement of mathematical skills in describing wave problems, but we have not considered the constitutive equations of viscoelastic media. Most of physical parameters can be roughly divided into three kinds, namely media properties, such as density, lame coefficients, and porosity; motion parameters, such as velocity, damping coefficient, and quality factor; state parameters, such as temperature and pressure. The coefficient of viscosity η is the most important physical parameter for describing the properties of viscoelastic media.

There are three kinds of basic viscoelastic models for rocks with different constitutive equations, and they are as follows (Christenson, 1982):

1. Voigt model;
2. Maxwell model;
3. Standard linear body model.

The second kind of model is seldom used in seismology. In the first kind of model, viscosity of the media is defined by viscosity coefficient η.

Assume that viscosity has just an effect on shear strain and that the viscous effect on wave propagation is proportional to η. Then, we can take $\mu + \eta \frac{d}{dt}$ instead of the shear modulus μ for the corresponding incident plane S-wave, and the wave Eqn (1.30) can be written as

$$\rho \frac{\partial^2 u}{\partial t^2} = \left(\mu + \eta \frac{d}{dt} \right) \Delta^2 \mu \tag{1.42}$$

In the Voigt model, what is the relationship between phase velocity c, attenuation coefficient α, and the viscosity coefficient η? Substituting new shear modulus $\mu + \eta \frac{d}{dt}$ into Eqn (1.39), we have

$$V(i\omega) = \sqrt{\frac{\mu + i\omega\eta}{\rho}} \tag{1.43}$$

Based on the Taylor's series expansion of $1/V(i\omega)$, the phase velocity c and attenuation coefficient α defined by Eqns (1.34) and (1.37) become

$$c(\omega) = \sqrt{\frac{2(\mu^2 + \omega^2\eta^2)}{\rho\mu + \rho\sqrt{(\mu^2 + \omega^2\eta^2)}}} \tag{1.44}$$

$$\alpha(\omega) = \omega\sqrt{\frac{\rho\sqrt{(\mu^2 + \omega^2\eta^2)} - \rho\mu}{2(\mu^2 + \omega^2\eta^2)}} \tag{1.45}$$

Therefore, the relationship between the coefficient of viscosity η and phase velocity c and attenuation coefficient α is fixed in the Voigt model.

In the Voigt model, we consider just the effect of the viscous coefficient on shear strain. If we further consider the effect of viscosity on dilatational strain as well, the description of viscoelastic model must be more accurate. The improved model is called the Kelvin–Voigt model. This model uses two coefficients of viscosity, η_1 and η_2, to construct new Lame coefficients as follows:

$$\lambda + \eta_1 \frac{\partial}{\partial t}; \ \lambda + \eta_2 \frac{\partial}{\partial t} \tag{1.46}$$

By applying the same method for deriving Eqn (1.42), we can get the wave equation corresponding to the Kelvin–Voigt model as follows:

$$\rho \frac{\partial^2 u}{\partial t^2} = (\lambda + 2\mu)\nabla^2 u + (\eta_1 + 2\eta_2) \frac{\partial}{\partial t} (\nabla^2 u) \tag{1.47}$$

Comparing Eqn (1.47) with Eqn (1.42) produced by the Voigt model, one finds that the second term is added on the right side of Eqn (1.47), which shows attenuation and dispersion of the P-wave in viscoelastic media.

1.6. ACOUSTIC WAVE EQUATION WITH VARIANT DENSITY OR VELOCITY

In the above discussion, we did not consider spatial variation of density and wave velocity in continuous media. We will take account of the wave motion in heterogeneous media as follows: As a constitutive equation in mechanics, Hooke's law resulted from laboratory experiments that used homogeneous media to perform experiments. Generalizing the constitutive equation to heterogeneous media must be conditional. In this section, we only consider the spatial variation of density or wave velocity in continuous media, and we assume that when density varies in space, wave velocity remains constant, or when wave velocity varies in the space, density is unchanged.

Assuming wave velocity to remain constant, when density varies in space, density $\rho(x)$ is a continuous function of spatial coordinate x. In this case, diffusion of wave energy is connected with $\rho(x)$. Thus, the wave diffusion factor is no longer the Laplace operator, but becomes $\nabla\left(\frac{1}{\rho}\nabla u\right)$ with a weight of $1/\rho$. For a point source, the variable density acoustic equation, corresponding to Eqn (1.19d), can be written as (Tiknonov and Samarskii, 1963; Du Shi-Tong, 2009)

$$\frac{1}{C^2}\frac{\partial^2 u}{\partial t^2} - \nabla\left(\frac{1}{\rho}\nabla u\right) = -\delta(\mathbf{x} - \mathbf{s})W(t) \tag{1.48}$$

where u denotes acoustic waves; c is the wave velocity; $\mathbf{s}$ is the location of the point source; and $W(t)$ is the wavelet function of the source.

In the 2D space, expanding the operator in Eqn (1.48), and using (x,z) to express a spatial point, we have

$$\left(\frac{\partial^2 u}{\partial x^2} + \frac{\partial^2 u}{\partial z^2}\right) - \frac{1}{C^2}\frac{\partial^2 u}{\partial t^2} - \frac{1}{\rho}\left(\frac{\partial\rho}{\partial x}\frac{\partial u}{\partial x} + \frac{\partial\rho}{\partial z}\frac{\partial u}{\partial z}\right) = -\delta(\mathbf{x} - \mathbf{s})W(t) \tag{1.49}$$

Equations (1.48) and (1.49) are the acoustic wave equations with varying densities of continuous media. In the Appendix of this book, we have introduced the finite difference method for the numerical solution of Eqn (1.49). To solve the acoustic wave equation, the initial-boundary conditions must be given.

Assuming that density remains unchanged when velocity changes, that is wave velocity $c(x)$ is not constant but is a continuous function of x. Then, the corresponding wave diffusion can no longer be the Laplace operator. According to Hooke's law, stress and strain are positively proportional, Lame coefficients are continuous functions $\lambda(x)$ and $\mu(x)$, $x = (x, y, z)$, Eqn (1.18) can be used as follows:

$$A_x = (\lambda + 2\mu)grad\ \mathbf{u}$$
$$A_y = (\lambda + 2\mu)grad\ \mathbf{v}$$
$$A_z = (\lambda + 2\mu)grad\ \mathbf{w}$$

In the continuous media, the divergence of stress is usually equivalent to the body force. In the absence of body force, substituting Eqn (1.18) into the equation of motion similar to Eqn (1.19) and considering variable velocity we have

$$\frac{\partial^2 u}{\partial t^2} = \frac{\partial}{\partial x}\left[c_p^2 \left(\frac{\partial u}{\partial x} + \frac{\partial v}{\partial y} + \frac{\partial w}{\partial z} \right) \right] \tag{1.50a}$$

$$\frac{\partial^2 v}{\partial t^2} = \frac{\partial}{\partial y}\left[c_p^2 \left(\frac{\partial u}{\partial x} + \frac{\partial v}{\partial y} + \frac{\partial w}{\partial z} \right) \right] \tag{1.50b}$$

$$\frac{\partial^2 w}{\partial t^2} = \frac{\partial}{\partial z}\left[c_p^2 \left(\frac{\partial u}{\partial x} + \frac{\partial v}{\partial y} + \frac{\partial w}{\partial z} \right) \right] \tag{1.50c}$$

Equations (1.50) are the acoustic wave equations with variant velocity, but they do not belong to typical acoustic wave equations jet, due to already coupling of the three wave components (u,v,w). If sedimentary basins are simplified as many horizontal formation layers, the wave velocity $c_p(x)$ can be approximately regarded as $c_p(x) = c_p(z)$. In this case, c_p can be extracted from a derivation of z. Considering the acceptable condition of nearly vertical reflection of single component w, namely $\frac{\partial w}{\partial z} \gg \frac{\partial w}{\partial x}$, $\frac{\partial w}{\partial z} \gg \frac{\partial w}{\partial y}$, Eqn (1.50c) turns approximately to acoustic Eqn (1.19d).

We have analyzed the condition of application of basic wave Eqn (1.19d) for single component reflection seismology. More attention should also be paid to the initial conditions and boundary conditions. The common initial condition for reflection seismology is

$$u\bigg|_{t=0} = 0, \quad \frac{\partial u}{\partial t}\bigg|_{t=0} = 0$$

Let u be the pressure or scalar displacement wave field. On an interface between two different homogeneous layers, the geometry of the interface is denoted as $z = f(x, y)$, the common boundary conditions are

Normal derivative is continuous : $\quad u_z^{(1)}\bigg|_{z=f(x,y)} = u_z^{(2)}\bigg|_{z=f(x,y)}$

Pressure or displacement is continuous : $\quad u^{(1)}\bigg|_{z=f(x,y)} = u^{(2)}\bigg|_{z=f(x,y)}$

Using the finite difference or finite element methods, numerical solutions of the initial and boundary value problem of the acoustic wave equation can be obtained. We recommend that postgraduate students majoring in applied geophysics gain practice on the subject as a course exercise (see Appendix A).

From the viewpoint of mathematics, the acoustic wave equation with variable velocity is much more complicated than ordinary initial-boundary value problems of common partial differential equations. The solutions are no longer ordinary functions but they belong to generalized functions. This topic will be further discussed in Chapters 3, 6, and 7. In the reflection seismology of a single component, the seismic wave is usually represented by ordinary functions, implying that some information is ignored. We should remember that a reflection seismic wave function is usually an approximation of real reflection seismic waves.

1.7. SUMMARY

This chapter focuses on the most basic concepts and operators of acoustic wave motion, such as continuous medium, the surface force, vibration, the description of the diffusion and wave propagation. If the medium is homogeneous and isotropic and the wave field is sufficiently far away from the power source, these equations can be applied to sound pressure wave fields, the pressure wave field in sea water, and also the potential of pure P-waves or S-waves in solids. In the following chapters, we will consider more complicated situations that cause more complex forms of wave equations. In addition, reflection seismology is mainly used for oil and gas exploration and engineering investigation, and effects of temperature and pressure changes on wave propagation can be ignored. This is why this book does not involve wave motion associated with heat flow or rheological mantle rocks.

The theory of particle dynamics founded in the age of Newton and Hooke revealed the connotation of vibration accurately. Later on, scientists, such as Huygens and Fermat, developed the geometric approximation on wave motion. The acoustic wave theory has been applied to industries gradually and successfully in Europe. However, seismic waves are propagating in continuous media within the Earth, and there exist some limitations of using particle dynamics to explain wave phenomenon in solids. In the twentieth century, the theory of continuum mechanics has been developed, which explains the connotation of the wave phenomenon in solids. In the next chapter, we will give an overview of wave continuum mechanics and discuss elastic wave propagation in solids.

CHAPTER

2

Elastic Waves in a Perfect Elastic Solid

Wave motion has been discussed in the previous chapter, and the acoustic wave equation for describing general wave motion has been derived. In this chapter, we will discuss elastic wave propagation in solids based on continuum mechanics. The acoustic wave equation is suitable for wave propagation in seawater, but it is not accurate in solid media that have tangential traction ability. In this chapter, we will discuss the elastic waves propagated in a perfectly elastic solid based on the theory of continuum mechanics (Fung, 1977; Spencer, 1980; Du Xun, 1985). Continuous media are composed of many fully connected grains, such as different minerals, and the product of the volume of a grain and its density is called the mass element. Forces are classified as surface force and body force, and we must use different expressions to show these forces. Surface force and body force should not be confused. Continuum mechanics starts from continuous media and continuous Eqn (1.2). The wave propagation in continuous media deals with mass grains and their surfaces, but not with free-motion particles in vibration.

2.1. STRESS TENSOR AND STRAIN TENSOR

Supporting a mass grain in continuous solid media is a cube with six faces; its volume is $dv = dx\,dy\,dz$ and its mass is $dm = \rho\,dv$. "Stress" refers to the reacting forces of a solid mass element in response to inserting external forces. "Strain" refers to the adjustment of the internal structures of a solid mass element under the action of the external forces. Explosions or vibrators produce strong body forces, causing media vibration and wave spreading. The body wave induces a surface force acting on the adjacent mass grains to generate displacement, which turns to produce strain all around. The strain acts on adjacent mass grains and enables them to produce traction stress. Finally, stress makes the adjacent mass grains to again generate acceleration and form an elastic wave. The elastic wave is just the spreading cycles of the stress–strain interaction in the far wavefield of reflection.

In continuous media, forces are classified into surface force and body force. Does elastic stress belong to body force or to surface force? Body force, such as gravitational force and electromagnetic force, is a distance force acting on all mass grains in continuous media. Surface force works directly on contact surfaces of grains, and its amplitude and direction depend on the surface area and normal direction. Water pressure, shear stress, and friction belong to surface force. For seismic wave propagation, elastic force exists between adjacent mass grains, so it does not belong to body force, but it belongs to surface force. In the far wavefield of reflection, the vibration of a mass grain is subjected to surface force and body force at the same time. However, due to the gravitational effect of a grain usually being far less than the surface force, we generally treat reflective waves as the surface force.

In solids, elastic stress acting on a mass grain includes the expansion–contraction force and the shearing force. A cubic mass grain has six faces; on each of the faces, the stress is different. The stress forced by the adjacent element is not necessarily perpendicular to the surface of a mass grain. Therefore, the stress is not represented by a vector, but by a second-order tensor that is denoted as A. Assume f_s to be the surface force acting on a surface, n to be the unit vector normal to the surface, and A to be the stress tensor of a mass grain. Then, we have

$$f_s = A \cdot n$$

Stress causes the deformation of mass grains, and the quantitative description of deformation is called strain, denoted as E, which is also a second-order tensor. In 3D space, the stress tensor along three axes can be expressed as

$$\mathbf{A} = (A_{ij}) = \begin{bmatrix} A_{11} & A_{12} & A_{13} \\ A_{21} & A_{22} & A_{23} \\ A_{31} & A_{32} & A_{33} \end{bmatrix} \tag{2.1}$$

The stress tensor has nine components, $A_{i,j}$, $i,j = 1,2,3$.

Can surface force and body force, being of different modes, get inter-converted? Assume that the density of a mass grain is ρ, its mass $m = \rho \Delta x \Delta y \Delta z$, and that its acceleration is a; then, its body force is f_b. On the other hand, no matter how to divide the mass grains, the surface force is $f_s = A \cdot n$. The body force acts to a unit volume dv. The surface force acts a unit surface ds. If we try to make them equivalent, we must add some additional conditions. That is, the surface integral of the surface force over the grain faces should be equal to the volume integral of the body force over the grain. This additional condition is also a necessary condition for stress equilibrium within continuous media. The unit normal vector shows that

$$\mathbf{n} = \alpha \mathbf{i} + \beta \mathbf{j} + \gamma \mathbf{k}; \quad \alpha^2 + \beta^2 + \gamma^2 = 1$$

[i,j,k] are unit vectors in [x,y,z] directions, the additional condition represents

$$\int f_s \, ds = \int A \cdot n \, ds = \int div \, \mathbf{A} \, dv \tag{2.2}$$

This equation shows the stress equilibrium in continuous media. Without Eqn (2.2), the motion equation in continuous media, which will be displayed later in Eqn (2.3), cannot be completed. Stress equilibrium means that the body force is equivalent to divergence of the stress tensor. In reflection seismology, stress is not in an equilibrium state only in a small neighborhood adjacent to shot points, whereas stress waves propagate in the equilibrium state in the far field. This book only deals with the cases of stress equilibrium in continuous media.

In continuous media, mass motion is caused by both the body force f_b and surface force $divA$, and substituting Eqn (2.2) into Newton's second law of motion yields

$$\rho(a - f_b) = div \, \mathbf{A} \tag{2.3a}$$

where a is the acceleration. Let us denote u as the displacement vector, v as the displacement velocity vector, and c as the wave velocity. Under the condition of $c \frac{\partial}{\partial x} \ll \frac{\partial}{\partial t}$, we have

$$v(x, t) = \frac{\partial u(x, t)}{\partial t}; \quad a(x, t) = \frac{\partial v(x, t)}{\partial t}$$

In reflection seismology, we do not consider the effect of the body wave f_b as wave propagation in the "far wavefield" is under consideration. Therefore, the motion equation in continuous media becomes (Fung, 1977; Spencer, 1980; Du Xun, 1985)

$$\rho \frac{\partial^2 \mathbf{u}}{\partial t^2} = div\ \mathbf{A} \tag{2.3b}$$

As the product of mass grain and density is called the mass element, by ignoring the instant change in the density we see that the mass element and mass grain are equivalent in motion. Mass grains undergo small deformations when suffering an elastic force during waves propagated in a solid, and the deformation includes instantaneous expansion or contraction and torsion, and is described by the strain tensor corresponding to the stress tensor. The strain tensor has nine components as in

$$\mathbf{E} = \left(E_{ij}\right) = \begin{bmatrix} E_{11} & E_{12} & E_{13} \\ E_{21} & E_{22} & E_{23} \\ E_{31} & E_{32} & E_{33} \end{bmatrix} \tag{2.4}$$

Denoting a point $(x,\ y,\ z) = (x_1,\ x_2,\ x_3)$, the relationship between the strain tensor and displacement is

$$E_{ij} = \frac{1}{2}\left(\frac{\partial u_j}{\partial x_i} + \frac{\partial u_i}{\partial x_j}\right); \quad i, j = 1, 2, 3 \tag{2.5}$$

In reflection seismology, we can simplify Hooke's law under some assumptions. First of all, we can assume that mass grains do not rotate in a wave field. Thus, the elastic stress tensor becomes a symmetric tensor, that is $A_{ij} = A_{ji}$ or $\mathbf{A} = \mathbf{A}^T$. As a result, the stress tensor and strain tensor have only six independent elements in this elastic media. In Cartesian coordinates, the stress tensor can be written as a vector as follows:

$$\mathbf{A} = \left(A_{xx}, A_{yy}, A_{zz}, A_{xy}, A_{yz}, A_{zx}\right)^T.$$

When mass grains do not rotate, the strain tensor also becomes a symmetric tensor and has six independent elements, written as $E_{ij} = E_{ji}$ or $E = E^T$. Thus, the strain tensor is reduced to a vector:

$$\mathbf{E} = \left(E_{xx}, E_{yy}, E_{zz}, E_{xy}, E_{yz}, E_{zx}\right)^T.$$

In isotropic media, gradients of displacement vector $u = (u,v,w)$ with respect to spatial directions (x,y,z) are proportional to the strain tensor with six components as follows:

$$E_{xx} = \left(\frac{\partial u}{\partial x}\right), \quad E_{yy} = \left(\frac{\partial v}{\partial y}\right), \quad E_{zz} = \left(\frac{\partial w}{\partial z}\right)$$

$$E_{xy} = E_{yx} = \frac{\partial v}{\partial x} + \frac{\partial u}{\partial y}$$

$$E_{yz} = E_{zy} = \frac{\partial w}{\partial y} + \frac{\partial v}{\partial z} \qquad (2.6a)$$

$$E_{zx} = E_{xz} = \frac{\partial u}{\partial z} + \frac{\partial w}{\partial x}$$

The first three parts of Eqn (2.6a) correspond to compression–dilation deformation, but the last three correspond to shear deformation. Based on Eqn (2.6a), we can see the divergence of the displacement vector:

$$div \, \mathbf{u} = E_{xx}\mathbf{i} + E_{yy}\mathbf{j} + E_{zz}\mathbf{k} \qquad (2.7)$$

where $\mathbf{k} = (\mathbf{i}, \mathbf{j}, \mathbf{k})$ are the unit directional vectors.

2.2.   VECTOR WAVE EQUATION IN FULLY ELASTIC MEDIA

After discussing the stress tensor and strain tensor, we are going to derive the vector wave equation according to the motion equation and constitutive equation. The wave field described by the vector wave equation can be observed and it results in three-component seismic records. In the next section, we will further discuss the scalar seismic wave equations corresponding to three-component seismic records.

The constitutive equation for linear elastic media is called the generalized Hooke's law. The elastic media that follow the generalized Hooke's law is called the linear elastic solid, which is usually accepted for cases of small deformations of a solid. For linear elastic media, the relationship between the stress tensor and strain tensor becomes (Eringen and Suhubi, 1975; Fung, 1977; Spencer, 1980; Du Xun, 1985)

$$A_{ij} = C_{ijkl}E_{kl} \qquad (2.8)$$

where $[C_{ijkl}]$ denotes a fourth-order tensor, called the elastic stiffness coefficient tensor. The fourth-order tensor has $3^4 = 81$ scalar elements. It is assumed that mass grains do not rotate in a wave field, and the stress tensor and strain tensor become symmetric tensors. As a result, the elastic stiffness tensor also becomes a symmetric tensor. For a linear elastic solid, we have

$$C_{ijkl} = C_{jikl} = C_{ijlk} = C_{klij} \qquad (2.9a)$$

This equation shows that for a linear elastic solid the tensor has only 21 independent elements, and the corresponding stress and strain tensors both have six elements. Thus, Eqn (2.8) can be simplified to

$$
\begin{bmatrix} A_{xx} \\ A_{yy} \\ A_{zz} \\ A_{xy} \\ A_{yz} \\ A_{zx} \end{bmatrix} =
\begin{bmatrix}
C_{11} & C_{12} & C_{13} & C_{14} & C_{15} & C_{16} \\
C_{12} & C_{22} & C_{23} & C_{24} & C_{25} & C_{26} \\
C_{13} & C_{23} & C_{33} & C_{34} & C_{35} & C_{36} \\
C_{14} & C_{24} & C_{34} & C_{44} & C_{45} & C_{46} \\
C_{15} & C_{25} & C_{35} & C_{45} & C_{55} & C_{56} \\
C_{16} & C_{26} & C_{36} & C_{46} & C_{56} & C_{66}
\end{bmatrix}
\begin{bmatrix} E_{xx} \\ E_{yy} \\ E_{zz} \\ E_{xy} \\ E_{yz} \\ E_{zx} \end{bmatrix}
\tag{2.9b}
$$

This equation is a linear matrix equation, representing the constitutive equation for an anisotropic elastic solid. For an isotropic linear elastic solid, Eqn (2.19) can be further simplified. Denote λ, μ are the Lame coefficients, due to

$$C_{12} = C_{11} - 2C_{44} = \lambda; \quad C_{11} = \lambda + 2\mu; \quad C_{44} = \mu$$

An isotropic elastic solid can be perfectly described by two elastic parameters to give the constitutive equation as follows:

$$
\begin{bmatrix} A_{xx} \\ A_{yy} \\ A_{zz} \\ A_{xy} \\ A_{yz} \\ A_{zx} \end{bmatrix} =
\begin{bmatrix}
\lambda + 2\mu & \lambda & \lambda & 0 & 0 & 0 \\
\lambda & \lambda + 2\mu & \lambda & 0 & 0 & 0 \\
\lambda & \lambda & \lambda + 2\mu & 0 & 0 & 0 \\
0 & 0 & 0 & \mu & 0 & 0 \\
0 & 0 & 0 & 0 & \mu & 0 \\
0 & 0 & 0 & 0 & 0 & \mu
\end{bmatrix}
\begin{bmatrix} E_{xx} \\ E_{yy} \\ E_{zz} \\ E_{xy} \\ E_{yz} \\ E_{zx} \end{bmatrix}
\tag{2.9c}
$$

For a completely isotropic elastic solid, the elastic stiffness tensor can be further simplified as follows:

$$C_{ijkl} = \lambda \delta_{ij} \delta_{kl} + \mu (\delta_{ik} \delta_{jl} + \delta_{il} \delta_{jk}) \tag{2.10}$$

Substituting Eqn (2.10) into Eqn (2.7), for an isotropic elastic solid, we rewrite the constitutive equation as

$$\mathbf{A} = \lambda (div \ \mathbf{u})\mathbf{I} + 2\mu \mathbf{E} \tag{2.11}$$

where $\mathbf{I}$ is the unit matrix. The corresponding equation of the components becomes

$$A_{ij} = \lambda E_{kk} \delta_{ij} + 2\mu E_{ij} \tag{2.12}$$

Ignoring the body force, and substituting Eqn (2.11) into Newton's motion Eqn (2.3b) yield

$$\rho \frac{\partial^2 \mathbf{u}}{\partial t^2} = div \ \mathbf{A} \tag{2.13}$$

referring to the formula in vector analysis (Pearson, 1974; Beyer, 1981)

$$\nabla^2 \mathbf{u} = grad \; div \; \mathbf{u} - curl \; curl \; \mathbf{u}$$

and with notation in (2.6a)

$$curl \; \mathbf{u} = \left(\frac{\partial w}{\partial y} - \frac{\partial v}{\partial z}, \frac{\partial u}{\partial z} - \frac{\partial w}{\partial x}, \frac{\partial v}{\partial x} - \frac{\partial u}{\partial y} \right)$$

$$\frac{\partial A_{ij}}{\partial x_i} = \lambda \frac{\partial}{\partial x_j} (div \; \mathbf{u}) + \mu \nabla^2 u_j$$

We substitute Eqn (2.11) into Eqn (2.13) and obtain

$$\rho \frac{\partial^2 \mathbf{u}}{\partial t^2} = (\lambda + \mu) grad(div \; \mathbf{u}) + \mu \nabla^2 \mathbf{u} \tag{2.14}$$

where $\mathbf{u}$ is the displacement vector and ∇^2 is the Laplace operator. This equation is the vector wave equation of displacement in an isotropic elastic solid, namely the Navier equation for elastic waves. The vector $\mathbf{u}$ in Eqn (2.14) contains longitudinal waves, shear waves, and converted waves, showing the complexity of wave propagation in a solid. Attention should be paid to the acoustic wave equations that were discussed in Chapter 1 and contain only longitudinal waves. Although Eqn (2.14) is more accurate for elastic solids, Eqn (2.14) is still not accurate enough for heterogeneous media. We will further discuss elastic wave equations with variable coefficients, which relates to the heterogeneous earth, in Section 3.1.

For a linear isotropic elastic solid, we can also use the potential of the displacement to express the displacement wave field (Budak et al., 1964; Telford et al., 1990) as

$$\mathbf{u} = grad \; \Delta + curl \; \mathbf{\Psi} \tag{2.15}$$

where Δ is the Lame potential or scalar potential, representing the potential of compressional waves and $\mathbf{\Psi}$ is the displacement potential of a shear wave. Be aware that it is a vector having three components. Note also that no rotation is assumed for mass grains in the solid; shear strain can also generate rotating waves such as surface waves, and so do not confuse rotary motion and rotational waves. Shear waves meet the condition

$$Div \; \mathbf{\Psi} = 0$$

Substituting Eqn (2.14) into Eqn (2.15) and expanding the operator yield

$$\rho \frac{\partial^2 \Delta}{\partial t^2} = (\lambda + 2\mu) \nabla^2 \Delta, \tag{2.16}$$

$$\rho \frac{\partial^2 \mathbf{\Psi}}{\partial t^2} = \mu \nabla^2 \mathbf{\Psi} \tag{2.17}$$

These two equations are displacement potential wave equations in completely isotropic elastic media. Among them, the Lame potential Δ represents

$$\Delta = E_{xx} + E_{yy} + E_{zz} = \frac{\partial u}{\partial x} + \frac{\partial v}{\partial y} + \frac{\partial w}{\partial z} \tag{2.18}$$

The Lame potential satisfies Eqn (2.16), which is the same as the acoustic wave Eqn (1.19). Equation (2.17) is the vector wave equation, because the potential of a shear wave cannot be expressed by a scalar. Equations (2.16) and (2.17) show the correlation between the Navier equation and the acoustic wave equation for body waves propagating in a solid. If the propagation of elastic waves in a solid could be decomposed into pure compressional waves and pure shear waves, then the pure-wave propagation would still follow the acoustic wave equation. The fourth chapter will discuss how to find pure compressional waves during seismic reflection processing.

In a one-dimensional case, the vector wave Eqn (2.14) is naturally simplified to a scalar wave equation. At this point, if the direction of wave propagation is the same as the vibration direction of a mass grain, for a compressional wave, Eqn (2.14) can be simplified as follows:

$$\frac{\partial^2 u}{\partial t^2} = \frac{\lambda + 2\mu}{\rho} \frac{\partial^2 u}{\partial x^2} \tag{2.19}$$

If the direction of wave propagation is perpendicular to the vibration direction of a mass grain, Eqn (2.14) can be simplified for a shear wave as follows:

$$\frac{\partial^2 u}{\partial t^2} = \frac{\mu}{\rho} \frac{\partial^2 u}{\partial x^2} \tag{2.20}$$

From these two equations, we can see that the corresponding P-wave and S-wave phase velocities equal

$$V_p = \sqrt{\frac{\lambda + 2\mu}{\rho}} \tag{2.21}$$

$$V_s = \sqrt{\frac{\mu}{\rho}} \tag{2.22}$$

respectively. However, the compressional and shear waves couple with each other in the elastic solid. Although in the infinite isotropic elastic solid there exist only compressional and shear waves, in limited elastic media, shear strain can produce rotating waves, such as surface waves, which we will discuss later.

2.3. SCALAR WAVE EQUATIONS IN FULLY ELASTIC MEDIA

The Navier equation is a very elegant representation, but it is not convenient for computation. Therefore, it is necessary to derive the corresponding scalar wave equations for computation. In this section, we will derive the explicit wave equations of displacement components equivalent to Navier Eqn (2.14), and it is assumed that mass grains do not rotate when vibrating. First of all, we can proceed from the potential function, and then move on to the expressions of displacement components.

The stress tensor in a linear elastic solid is a symmetric tensor; stress and strain tensors have only six independent elements, as shown in Eqn (2.7). In the condition of three-dimensional linear isotropic elastic media, we can use the Lame coefficient (λ, μ) to describe the elastic parameters of a solid. To express three-component seismic records, let the three displacement components with respect to (x,y,z) be $(u,v,w)^T$. The relationship between the six components of strain, and the displacement is

$$E_{xx} = \frac{\partial u}{\partial x}, \quad E_{yy} = \frac{\partial v}{\partial y}, \quad E_{zz} = \frac{\partial w}{\partial z}$$

For a normal strain component,

$$E_{xy} = E_{yx} = \frac{\partial v}{\partial x} + \frac{\partial u}{\partial y}$$

For a tangential strain component,

$$E_{yz} = E_{zy} = \frac{\partial w}{\partial y} + \frac{\partial v}{\partial z} \quad E_{zx} = E_{xz} = \frac{\partial u}{\partial z} + \frac{\partial w}{\partial x} \quad (2.6b)$$

Let the density of a mass grain be ρ, and let its mass equal $m = \rho \triangle x \triangle y \triangle z$. Substituting Eqn (2.7) into Eqn (2.14) yields

$$\rho \frac{\partial^2 u}{\partial t^2} = (\lambda + \mu) \frac{\partial \Delta}{\partial x} + \mu \nabla^2 u, \quad (2.23a)$$

$$\rho \frac{\partial^2 v}{\partial t^2} = (\lambda + \mu) \frac{\partial \Delta}{\partial y} + \mu \nabla^2 v, \quad (2.23b)$$

$$\rho \frac{\partial^2 w}{\partial t^2} = (\lambda + \mu) \frac{\partial \Delta}{\partial z} + \mu \nabla^2 w, \quad (2.23c)$$

Among which

$$\nabla^2 u = (\partial^2 u/\partial x^2 + \partial^2 u/\partial y^2 + \partial^2 u/\partial z^2), \quad (2.24)$$

After taking differential operation to Eqn (2.16) and then addition of all terms result in

$$\rho \frac{\partial^2}{\partial t^2} \left(\frac{\partial u}{\partial x} + \frac{\partial v}{\partial y} + \frac{\partial w}{\partial z} \right) = (\lambda + \mu) \left(\frac{\partial^2 \Delta}{\partial x^2} + \frac{\partial^2 \Delta}{\partial y^2} + \frac{\partial^2 \Delta}{\partial z^2} \right)$$
$$+ \mu \nabla^2 \left(\frac{\partial u}{\partial x} + \frac{\partial v}{\partial y} + \frac{\partial w}{\partial z} \right),$$

If we substitute the above equation into Eqn (2.18), we obtain a compressive wave equation specially for the Lame potential, the same as Eqn (2.16). Similarly, we can obtain the vector potential equations of a pure shear wave to be the same as Eqn (2.17). For example, the pure shear wave propagating perpendicularly to the x-direction in (y, z) plane is

$$\rho \frac{\partial^2}{\partial t^2} \left(\frac{\partial w}{\partial y} - \frac{\partial v}{\partial z} \right) = \mu \nabla^2 \left(\frac{\partial w}{\partial y} - \frac{\partial v}{\partial z} \right),$$

Therefore, the vector potential component of a shear wave propagating in the (y, z) plane is

$$\Psi_x = \left[\frac{\partial w}{\partial y} - \frac{\partial v}{\partial z} \right]$$

Similarly, the vector potential component of a shear wave propagating on the (x, z) plane is

$$\Psi_y = \left[\frac{\partial u}{\partial z} - \frac{\partial w}{\partial x} \right]$$

On the (x, y) plane,

$$\Psi_z = \left[\frac{\partial v}{\partial x} - \frac{\partial u}{\partial y} \right]$$

When we have vector potential components, it is not difficult to get the displacement potential vector. We should note that the three components of these scalar equations are coupled to each other, and it is impossible to decouple them. The three components of an elastic displacement wave field contain compressive waves, shear waves, and converted multiples, rather than perfectly pure waves. If one wants to convert the three-component recording into pure-wave recording, one must go through special seismic data processing.

We are looking for the scalar wave equations for the three wave field components of displacement, and using them to express the three-component seismic records. In a linear isotropic elastic solid, stress A and strain E have six axial components. If we substitute Eqn (2.9c) into

Hooke's law Eqn (2.12), we have the constitutive equation in scalar forms as (Eringen and Suhubi, 1975; Du Xun, 1985) in the following:

$$
\begin{aligned}
A_{xx} &= (\lambda + 2\mu)E_{xx} + \lambda E_{yy} + \lambda E_{zz} \\
A_{yy} &= \lambda E_{xx} + (\lambda + 2\mu)E_{yy} + \lambda E_{zz} \\
A_{zz} &= \lambda E_{xx} + \lambda E_{yy} + (\lambda + 2\mu)E_{zz} \\
A_{xy} &= \mu E_{xy} \quad A_{yz} = \mu E_{xy} \quad A_{zx} = \mu E_{zx}
\end{aligned}
\tag{2.25}
$$

These represent the surface force normal to the six surfaces of a mass grain. From Eqn (2.2), we know that the divergence of the stress in the six surfaces and body force are physically equivalent, and then the Lame potential is equivalent to the addition of coaxial strain components as follows:

$$
\Delta = E_{xx} + E_{yy} + E_{zz} = \frac{\partial u}{\partial x} + \frac{\partial v}{\partial y} + \frac{\partial w}{\partial z}
\tag{2.18}
$$

After the stress and strain are decomposed, substituting Eqns (2.7) and (2.25) into Eqn (2.2) yields

$$
\begin{aligned}
\rho \frac{\partial^2 u}{\partial t^2} &= (\lambda + 2\mu)\frac{\partial^2 u}{\partial x^2} + \mu\left(\frac{\partial^2 u}{\partial y^2} + \frac{\partial^2 u}{\partial z^2}\right) + (\lambda + \mu)\left(\frac{\partial^2 v}{\partial x \partial y} + \frac{\partial^2 w}{\partial x \partial z}\right) \\
\rho \frac{\partial^2 v}{\partial t^2} &= (\lambda + 2\mu)\frac{\partial^2 v}{\partial y^2} + \mu\left(\frac{\partial^2 v}{\partial x^2} + \frac{\partial^2 v}{\partial z^2}\right) + (\lambda + \mu)\left(\frac{\partial^2 u}{\partial x \partial y} + \frac{\partial^2 w}{\partial y \partial z}\right) \\
\rho \frac{\partial^2 w}{\partial t^2} &= (\lambda + 2\mu)\frac{\partial^2 w}{\partial z^2} + \mu\left(\frac{\partial^2 w}{\partial x^2} + \frac{\partial^2 w}{\partial y^2}\right) + (\lambda + \mu)\left(\frac{\partial^2 v}{\partial z \partial y} + \frac{\partial^2 u}{\partial x \partial z}\right)
\end{aligned}
\tag{2.26}
$$

In a homogeneous isotropic elastic solid, Eqn (2.26) are the scalar elastic wave equations that are suitable for the three-component reflection seismic records. The three terms on the right side of Eqn (2.26) have a clear physical meaning, representing compressive wave, shear wave, and converted wave, respectively. Although the three components in Eqn (2.26) are coupled to each other, and cannot be decoupled directly, in special cases Eqn (2.26) can be further simplified.

Because engineering geophysical surveys are usually carried out in soil, the shear stress becomes weaker in unconsolidated and water-saturated soil and S-wave propagation can be ignored, that is $\mu \ll \lambda$. If we ignore the shear stress, Eqn (2.25) can be simplified as

$$
\begin{aligned}
A_1 &= A_{xx} = (\lambda + 2\mu)E_1 + \lambda E_2 + \lambda E_3 \\
A_2 &= A_{yy} = (\lambda + 2\mu)E_2 + \lambda E_1 + \lambda E_3 \\
A_3 &= A_{zz} = (\lambda + 2\mu)E_3 + \lambda E_2 + \lambda E_1 \\
A_{xy} &= A_{yz} = A_{zx} = 0
\end{aligned}
\tag{2.27}
$$

Here, the stress has only three components in the directions (x,y,z), and are denoted as (A_1,A_2,A_3). We can also express strain with three axial components

$$E_1 = E_{xx}, \quad E_2 = E_{yy}, \quad E_3 = E_{zz}$$

In turn, the corresponding three axial components of strain can be expressed as

$$E_1 = A_1/Y - \sigma A_2/Y - \sigma A_3/Y$$
$$E_2 = A_2/Y - \sigma A_1/Y - \sigma A_3/Y \qquad (2.28)$$
$$E_3 = A_3/Y - \sigma A_2/Y - \sigma A_1/Y$$

where Y is Young's modulus and σ is Poisson's ratio. We will discuss Poisson's ratio and Young's modulus in Section 2.7. The three main components of strain $\{E_i\}$ are equal to the main strain (A_i/Y) minus other two orthogonal shear strains $(\sigma A_j/Y)$ with a weight of σ. Furthermore, the 3D three-component elastic wave equations can be simplified as

$$\rho \frac{\partial^2 u}{\partial t^2} = (\lambda + 2\mu)\left(\frac{\partial^2 u}{\partial x^2} + \frac{\partial^2 v}{\partial x \partial y} + \frac{\partial^2 w}{\partial x \partial z}\right) \qquad (2.29)$$

$$\rho \frac{\partial^2 v}{\partial t^2} = (\lambda + 2\mu)\left(\frac{\partial^2 v}{\partial y^2} + \frac{\partial^2 u}{\partial x \partial y} + \frac{\partial^2 w}{\partial y \partial z}\right) \qquad (2.30)$$

$$\rho \frac{\partial^2 w}{\partial t^2} = (\lambda + 2\mu)\left(\frac{\partial^2 w}{\partial z^2} + \frac{\partial^2 v}{\partial z \partial y} + \frac{\partial^2 u}{\partial x \partial z}\right) \qquad (2.31)$$

These equations show that compressional waves in a wave field dominate in soils and can be called compressional wave-dominated elastic equations. However, the three components of the wave field are also coupled.

2.4. ELASTIC WAVE EQUATION IN TWO-DIMENSIONAL MEDIA

In seismic profiling, two-dimensional (2D) seismic gathers are often used for regional surveys. Because only 2D profiling data are observed on the surface, it is impossible to get information of wave propagation outside the profile. It is necessary to study 2D stress fields and corresponding elastic wave equations. Especially in 2D surveys, a reflection wave field is usually expressed by a scalar wave equation similar to the acoustic wave equation. However, a mathematical analysis should be given for this simplified situation.

Denote the striking direction of layers as y, and a point in 2D profiles by (x, z). Based on the following four kinds of assumptions, a 2D elastic wave field can be simplified from a 3D elastic wave field:

1. Assume the stress in the y-direction equals zero: $A_{ij} = 0$.
2. Assume the strain in the y-direction equals zero: $E_{ij} = 0$.
3. Assume the stress does not change in the y-direction: $\frac{\partial A_{ij}}{\partial y} = 0$.
4. Assume the strain and two components of the displacement in the y-direction equals zero: $\frac{\partial u}{\partial y} = 0$, $\frac{\partial w}{\partial y} = 0$.

Because the first three assumptions are not in conformity with the situation of seismic observation, they are not adopted by reflection seismology. Substituting the fourth assumption into Eqn (2.26) yields

$$\rho \frac{\partial^2 u}{\partial t^2} = (\lambda + 2\mu) \frac{\partial^2 u}{\partial x^2} + \mu \frac{\partial^2 u}{\partial z^2} + (\lambda + \mu) \frac{\partial^2 w}{\partial x \partial z} \qquad (2.32)$$

$$\rho \frac{\partial^2 w}{\partial t^2} = (\lambda + 2\mu) \frac{\partial^2 w}{\partial z^2} + \mu \frac{\partial^2 w}{\partial x^2} + (\lambda + \mu) \frac{\partial^2 u}{\partial x \partial z} \qquad (2.33)$$

These belong to the coupled 2D wave equations of two components along the profile; each equation belongs to the acoustic wave equation, but with a coupling term that is corresponding to plane shear deformation. This set of equations can be used for the synthesis of a seismic wave field for two-component seismic profiling, but it is only accurate in isotropic media.

The simplifying process of the 2D seismic wave field is equivalent to decouple a y-direction wave field from the 3D wave field, that is decoupling the P/SV and SH waves. After the decoupling process, the wave component in the y-direction, that is $v(x, z, t)$, approximately satisfies the acoustic wave equation. In fact, it is necessary to consider the wave propagation in the anisotropic media in seismic data processing, where waves in the y-direction must be accurately described, and will be discussed as follows.

2.5. ELASTIC WAVE EQUATIONS IN ANISOTROPIC MEDIA

It has been discussed in Section 2.2 that the linear elastic medium satisfies Hooke's law and can be applied to describe a solid under conditions of small deformations. At constant temperature, both the stress and strain tensors are second-order symmetric tensors, and the stress components are linear functions of the corresponding strain functions as shown in Eqn (2.8):

$$A_{ij} = C_{ijkl} E_{kl} \qquad (2.34)$$

where the coefficient C_{ijkl} is the fourth-order tensor of elastic stiffness coefficients, and there are $3^4 = 81$ scalar elements. Assuming that a mass grain does not rotate during vibration, the elastic stiffness tensor is also a symmetric tensor, as shown in Eqn (2.9):

$$C_{ijkl} = C_{jikl} = C_{ijlk} = C_{klij} \qquad (2.35)$$

The elastic tensor in linear elastic media has only 21 independent elements, which is sufficient for typical anisotropic media. Only isotropic linear elastic media can be expressed by using two elastic parameters, that is the Lame coefficients. For anisotropic elastic media, at least more than three parameters should be used to describe their elastic properties because both compressive and shear deformations of a mass grain may have different directions.

Crystalline rocks are polymers of minerals. If the minerals are cubic, such as basalt and eclogite, then crystalline rocks can be treated as isotropic. Otherwise, it must belong to the anisotropic category, such as gneiss and rocks containing a lot of mica or feldspar. Sedimentary rocks contain mineral fragments and are porous. If they do not have dominant directions, in general, then they can be treated as isotropic, such as sandstone and limestone. Otherwise, they could be anisotropic, such as slate and shale. However, the most sedimentary rocks are stratified. If a sedimentary layer is very thin, the average elasticity of the multilayer sedimentary rocks will also be different with respect to different directions, causing macroscopic anisotropy. Therefore, its more accurate models have been considering characteristics of anisotropy in reflection seismology (Sheriff and Geldart, 1983; Wu and Valerie, 2007). Hereafter, we take nearly horizontal strata models as examples of a sedimentary basin. The elastic stiffness increases vertically due to the gravitational compaction of the strata, and the elasticity coefficients in the horizontal and vertical directions should be different. However, the elasticity coefficients in both the x-direction and the y-direction are the same. These media belong to the transversely isotropic media that are symmetric to the vertical axis, called vertically symmetry transverse isotropic (VTI) media (Eringen and Suhubi, 1975; Du Shi-Tong, 2009). The elastic coefficients of the VTI media can be described using five parameters as follows:

λ_h—strain ratio in the direction of horizontal primary stress
λ_v—strain ratio in the direction of vertical primary stress
u_r—converted shear strain ratio in the direction of the primary stress
u_h—horizontal shear strain ratio in a direction different from that of the primary stress

u_v—vertical shear strain ratio in a direction different from that of the primary stress. Hooke's law of the linear anisotropic media is written as

$$A_{xx} = \lambda_h E_{xx} + (\lambda_h - 2\mu)E_{yy} + \mu_r E_{zz}$$
$$A_{yy} = (\lambda_h - 2\mu_h)E_{xx} + \lambda_h E_{yy} + \mu_r E_{zz}$$
$$A_{zz} = \mu_r E_{xx} + \mu_r E_{yy} + \lambda_v E_{zz} \tag{2.36}$$
$$A_{xy} = \mu_h E_{xy}, A_{yz} = \mu_v E_{yz}, A_{zx} = \mu_v E_{zx}$$

By substituting the three displacement component expressions and Eqn (2.36) into the equation of motion we can obtain the elastic wave equation of the three components for VTI media as follows:

$$\rho \frac{\partial^2 u}{\partial t^2} = \lambda_h \frac{\partial^2 u}{\partial x^2} + \mu_h \frac{\partial^2 u}{\partial y^2} + \mu_x \frac{\partial^2 u}{\partial z^2} + (\lambda_h - \mu_h)\frac{\partial^2 v}{\partial x \partial y} + (\mu_r + \mu_x)\frac{\partial^2 w}{\partial x \partial z}$$

$$\rho \frac{\partial^2 v}{\partial t^2} = (\lambda_h - \mu_h)\frac{\partial^2 u}{\partial x \partial y} + \mu_h \frac{\partial^2 v}{\partial x^2} + \lambda \frac{\partial^2 v}{\partial y^2} + \mu_v \frac{\partial^2 v}{\partial z^2} + (\mu_r + \mu_v)\frac{\partial^2 w}{\partial y \partial z} \tag{2.37}$$

$$\rho \frac{\partial^2 w}{\partial t^2} = (\mu_r + \mu_h)\frac{\partial^2 u}{\partial x \partial z} + (\mu_r + \mu_v)\frac{\partial^2 v}{\partial z \partial y} + \mu_v \frac{\partial^2 w}{\partial x^2} + \mu_v \frac{\partial^2 w}{\partial y^2} + \lambda_v \frac{\partial^2 w}{\partial z^2}$$

By comparing Eqn (2.37) with the three-component elastic wave Eqn (2.26) in isotropic media, we can find the different features of waves in anisotropic media, which are caused mainly by adding terms of shear strain ratios.

The velocity of wave propagation in anisotropic media is dependent on the propagation direction. After Eqn (2.37) is simplified into a one-dimensional wave equation in VTI media by setting two displacement components to zero, we can obtain the velocities of P-waves and S-waves as follows:

$$\text{Horizontal } P - \text{wave velocity } V_{PH} = (\lambda_h/\rho)^{1/2}$$

$$\text{Vertical } P - \text{wave velocity } V_{PV} = (\lambda_v/\rho)^{1/2}$$

$$\text{Vertical } SV - \text{wave velocity } V_{sv} = (\mu_v/\rho)^{1/2}$$

$$\text{Horizontal } SH - \text{wave velocity } V_{sh} = (\mu_h/\rho)^{1/2}$$

There exist a fast shear wave and a slow shear wave in VTI media, and their velocities are denoted as V_{sv} and V_{sh}, respectively. The existence of the fast and slow shear waves has been proved by a large number of observations. The coaction of velocity variation in different directions on medium vibration is called polarization.

This VTI media model with five anisotropy parameters can be easily extended to the situation where anisotropies also exist in the horizontal directions, say x and y. In this case, the velocities of pure P-waves are different in the three directions and are controlled by proportional coefficients C_{11}, C_{22}, and C_{33}. The velocities of pure S-waves are different too and controlled by C_{44}, C_{55}, and C_{66}. The total number of elastic parameters reaches nine including the parameters of shear−strain conversion, say C_{12}, C_{13}, and C_{23}, which are symmetric along the three axes. If we refer to Eqn (2.36), the Hooke's law in the linear anisotropic media can be written as follows:

$$A_{xx} = C_{11}E_{xx} + C_{12}E_{yy} + C_{13}E_{zz}$$
$$A_{yy} = C_{12}E_{xx} + C_{22}E_{yy} + C_{23}E_{zz}$$
$$A_{zz} = C_{13}E_{xx} + C_{23}E_{yy} + C_{33}E_{zz} \tag{2.38}$$
$$A_{xy} = C_{44}E_{xy}, A_{yz} = C_{55}E_{yz}, A_{zx} = C_{66}E_{zx}$$

The nine-parameter anisotropic media is equivalent to the rock composed of hexagonal crystal minerals, or the homogeneous rocks containing three sets of orthogonal cracks. By substituting Eqn (2.38) into the equation of motion, one can derive the elastic wave equations in the anisotropic media with nine elastic parameters.

Theoretically, the analysis of elastic waves in anisotropic media is a very good subject for research, in which models can be changed slightly to improve the interpretation of three-component seismic records. For example, the interpretation of the horizontal components of seismic records cannot be done without the guidance of wave theory in anisotropic media.

2.6. BOUNDARY CONDITIONS FOR ELASTIC WAVE EQUATIONS

In the above, we have discussed the theory of elastic waves in an ideal elastic solid without a discussion on boundaries. However, the earth does have boundaries, and the existence of interfaces will cause some changes in vibration propagation and generate new seismic wave phases. Boundaries can be classified into three categories: solid/solid boundary, fluid/fluid boundary, and fluid/solid boundary. For seismic waves, we assume strata interfaces belonging to solid/solid boundaries, the surface of land and walls of drilling wells belonging to the fluid/solid boundaries, and the ocean's surface as a fluid/fluid boundary. Different types of waves correlate with different boundaries, for example, the Stoneley wave is very strong on the wall of a drilling well, and the surface wave is a specific type of wave that occurs at fluid/solid boundaries.

Generally, elastic boundary conditions can be classified into three categories as follows:

1. stress boundary condition;
2. displacement boundary condition;
3. mixed force (including body force and the surface force) and displacement boundary condition.

In fluids, there is no shear stress and thus no shear deformation. This is the main reason why fluid/solid interfaces have sharp discontinuities of stress and displacement.

In continuous media, a boundary can be regarded as a combination of many surface grain cells. If an interface is located at plane $z = 0$, the stress (surface force) acting up and down the boundaries of the interface mass grains is the same. Displacement of the mass grain may be continuous but the motion style can be different. In this section, we use square brackets "[*]" to represent the differences in stress or displacement components between the opposite sides of mass grains on the boundary. For a fluid/ fluid boundary, as $\mu = 0$, only the vertical components of the stresses and displacements on the opposite sides of the boundary are continuous, namely

$$[A_{zz}] = 0, \ [w] = 0 \tag{2.39}$$

For a solid/solid boundary, all the components of the stresses and displacements on the two sides of the boundary are continuous, namely

$$[A_{zx}] = 0, \ [A_{zy}] = 0, \ [A_{zz}] = 0;$$
$$[u] = 0, \ [v] = 0, \ [w] = 0 \tag{2.40}$$

Solid/solid boundaries generate reflective seismic waves, which are the most important signals we require.

For a fluid/solid boundary, due to the lost shear displacement in fluids, the horizontal components of the displacements on the two sides of the boundary are not continuous and the vertical components of stress and displacements are continuous

$$[A_{zx}] = 0, \ [A_{zy}] = 0, \ [A_{zz}] = 0, \ [w] = 0 \tag{2.41}$$

The fluid/solid boundaries, such as at the sea bottom, generate multiple seismic waves, which become boring noises in reflection seismology for offshore oil exploration. From the equations mentioned above, we know that the stress condition is generally applicable while the displacement condition varies according to the different types of boundaries.

The discontinuity of both the horizontal components of stress and displacement on the fluid/solid boundaries also generates surface waves propagating along the surface: the Rayleigh wave and Love wave. The Rayleigh wave is a surface wave that propagates along the surface of a semiinfinite elastic solid. Along the Rayleigh wave propagation, the vibration trajectory of a mass grain is elliptic with its long axis pointing to the z direction if the surface locating at plane $z = 0$. The orientation of the elliptic plane is the same as the direction of wave propagation. Thus, the Rayleigh wave belongs to SV-waves, and its velocity is about $0.92V_s$.

If there exists a thin and low-velocity layer on the surface of a semi-infinite elastic solid, that is a soil layer or the crust, a Love wave is generated and it belongs to SH surface waves. Its velocity has dispersion characteristics with frequencies. Assuming that the velocity of a surface S-wave is V_{s1} and that the S-wave velocity of the thin layer beneath is V_{s2} (here $V_{s2} > V_{s1}$). Then, the base order or small-periodic velocity of the Love wave approaches V_{s1}, and lager-periodic velocity of Love wave approaches V_{s2}.

For a solid/solid, all the vertical components of the displacement and stress on the boundary are continuous, and the interface generates the most secondary wave phases. If in the 2D case a planar interface is located at $z = 0$, the solid/solid boundary conditions Eqn (2.40) can be converted into the following formulas:

$$[A_{zx}] = \left[2\mu \frac{\partial^2 \Delta}{\partial x \partial z} + \mu \left(\frac{\partial^2 \Psi}{\partial x^2} - \frac{\partial^2 \Psi}{\partial z^2} \right) \right] = 0,$$

$$[A_{zz}] = \left[\lambda \nabla^2 \Delta + 2\mu \left(\frac{\partial^2 \Delta}{\partial z^2} - \frac{\partial^2 \Psi}{\partial x \partial z} \right) \right] = 0;$$

$$[u] = \left[\frac{\partial \Delta}{\partial x} - \frac{\partial \Psi}{\partial z} \right] \approx 0,$$

$$[w] = \left[\frac{\partial \Delta}{\partial z} + \frac{\partial \Psi}{\partial x} \right] \approx 0$$

(2.42)

The incident waves on the interface can be a downgoing P-wave, or an upgoing P-wave, or a downgoing SV-wave, and an upgoing SV-wave, and each of them produces from the interface secondary downgoing wave, and upgoing wave that means reflective waves. Therefore, there are totally 16 kinds of wave phases at the planar interface, including the four kinds of incident waves.

From the examples of Rayleigh waves and Love waves, we can see the significance of the boundaries to wave propagation in an elastic solid. Any boundary can cause the splitting of incident wave energy and can produce secondary waves with new phases. For example, an incident wave on the interface between two layers can produce not only the transmitted wave, reflective wave, converted wave but also the refracted wave propagating along the interface. A new boundary produces new wave phases propagating with different velocities, providing the basic principle for reflection seismology.

2.7. ELASTIC WAVE VELOCITIES OF ROCKS

It is necessary to discuss the rock physical parameters together with elastic waves propagating in a solid, because rock physics is also the application basis of reflection seismology in addition to elasticity mechanics. Suppose that V_p is the P-wave velocity, V_s is the S-wave velocity, σ is Poisson's ratio, according to the elastic wave Eqns (2.21) and (2.22), we know

$$V_s = \sqrt{\mu/\rho} \tag{2.43}$$

$$V_p = \sqrt{\frac{k + 4\mu/3}{\rho}} = \sqrt{\frac{\lambda + 2\mu}{\rho}} \tag{2.44}$$

$$\sigma = \frac{3k - 3\mu}{2(3k + \mu)} = \frac{\lambda}{2(\lambda + \mu)} \tag{2.45}$$

where k indicates the compression bulk modulus, μ is the shear modulus, λ indicates the Lame coefficient, and ρ is the density. Their relationship can be expressed as follows:

$$\lambda = \rho \left[V_P^2 - \frac{4}{3} V_S^2 \right] \tag{2.46}$$

$$\mu = \rho V_s^2 \tag{2.47}$$

Rocks are the multiphase media of polymineral mixtures. Wave propagation properties of Minerals, including P-wave velocities, shear moduli, compressive bulk moduli, and Poisson's ratios, are shown in Table 2.1 (Sheriff and Geldart, 1983; Mavco et al., 2009). P-wave and S-wave velocities of some general minerals increase with increasing contents of iron and magnesium in minerals, which is similar to the trend of increasing iron and magnesium contents with the increasing depth in the lithosphere.

The definition of Poisson's ratio Eqn (2.45) can also be rewritten as follows:

$$\sigma = 0.5 \left\{ 1 - 1 / \left[(V_p/V_s)^2 - 1 \right] \right\} = 0.5 \left[(V_p/V_s)^2 - 2 \right] / \left[(V_p/V_s)^2 - 1 \right] \tag{2.48}$$

or

$$\sigma = 0.5 \left[1 - 2(V_s/V_p)^2 \right] / \left[1 - (V_s/V_p)^2 \right] \tag{2.49}$$

TABLE 2.1   Wave Propagation Properties of Minerals (Room Temperature)

Mineral	Density (g/cm³)	V_p (km/s)	V_s (km/s)	k (10^{10} Pa)	μ	Poisson Ratio
Air	0.0012	0.343				
Vapor	0.0048	0.401				
Oil	0.7	1.33				
Water	0.96	1.43				
Glass	6.0	3.49		2.3	2.3	0.25
Steel	7.9	5.00		16.4	7.57	0.30
Copper	9.2	4.65	3.5	16.1	4.6	0.37
Gold	2.0	1.18		16.9	2.85	0.42
Lead	11.7	2.22	1.49	3.6	0.54	0.43
Quartz	2.65	5.44–6.03	4.11	4.4	0.08	
Melting quartz				3.7	3.12	0.17
Muscovite	2.79	5.81	3.39	3.07	0.28	0.24
Biotite	2.75	5.13	2.98	2.74	0.25	0.25
Orthoclase	2.57	5.9	3.07	2.42	0.31	0.31
Plagioclase	2.61	6.06	3.35	2.93	0.28	0.28
Diamond	3.5–3.6	16–18				
Olivine	3.32	8.42–8.57	4.89–5.16	8.08	0.24	0.22–0.25
Spinel	3.6	8.36–10	4.52–5.68	11.65	0.26	0.26–0.29
Orthopyroxene	3.16	7.93–8.08	4.74–4.87			0.21–0.22
Clinopyroxene	3.2–3.7	7.66–7.84	4.35–4.51			0.25–0.26
Essonite	3.6–4.18	9.31	5.43			0.24
Pyrope		8.96	5.05			0.27

We can see that Poisson's ratio does not change when the velocities of both *P*-waves and *S*-waves are changing proportionally, and it depends on the ratio between the velocities of *P*-waves and *S*-waves. Therefore, Poisson's ratio should be treated as an independent elastic parameter.

Theoretically, Eqn (2.45) can be written as

$$6 = (1 - \mu/k)/[2 + 0.333\,\mu/k] \qquad (2.50)$$

It is shown in Eqn (2.50) that Poisson's ratio is a nonlinear function of the ratio between shear elasticity and bulk compressibility, and it makes sense only for the compressible materials with $k > 0$. Supposing $\mu = 0$, for example in the ideal fluid, substituting it into Eqn (2.50) yields

$$V_s/V_p = 0; \quad 6 = 0.5 \tag{2.51}$$

Supposing $\mu = \lambda$, for example in the rigid solid, substituting it into Eqn (2.50) yields

$$V_s/V_p = 1/\sqrt{3}; \quad 6 = 0.25 \tag{2.52}$$

Supposing $\mu = k$, for example in rubbers, substituting it into Eqn (2.50) yields

$$V_s/V_p = 3/7 = 0.4286; \quad 6 = 0.0 \tag{2.53}$$

Therefore, for the solids of a uniform component, some substance can bear strong pressure and strong shear force at the same time, and their Poisson's ratios are in the range of 0.2–0.3. Another can bear strong pressure but weak shear force, and their Poisson's ratios decrease in comparison to that of the former and are close to 0.1–0.2. Others can bear only weak pressure and shear force at the same time, and their Poisson's ratios may increase to 0.43, for example the coal. For water-saturated sedimentary rock, λ is much greater than μ, so $6 > 0.25$, and it even reaches 0.4. Among natural state rocks, the Poisson's ratios of coal and shale are the highest. While Poisson's ratios of sandstone saturated with natural gas and carbonate reservoir rocks, as well as granite, are <0.2, due to their bulk compressibility, k increases.

Poisson's ratio of crystalline rocks in situ can be calibrated through measuring drill cores. Figure 2.1 shows the relationship between Poisson's ratio, ratio of P-wave velocity versus S-wave velocity, and P-wave velocity under different pressures. We can conclude that the Poisson's ratios of felsic rocks with the lowest velocity of P-waves are the lowest, say 0.22–0.27. These rocks forming the upper crust, for example sandstone and granite have low P-wave velocities. Poisson's ratio of mafic rocks is in the middle range, that is 0.25–0.30; these rocks form the lower crust and lithospheric mantle. Poisson's ratio of anorthosite rocks that form the middle crust is the highest, that is 0.29–0.32. Table 2.2 also shows the Seismic properties of crystalline rocks. The velocity and Poisson's ratio are measured from the cores of Chinese *continental* scientific drilling holes (Ji et al., 2002, 2007; Yang et al., 2008). The table shows that the Poisson's ratios of eclogite and basalt are the highest, the ratio of peridotite as the most popular mantle rock is around 0.25, and the ratios of gneiss and granite obviously decrease. Therefore, increasing Poisson's ratios become an indicator of the distribution of mantle rocks.

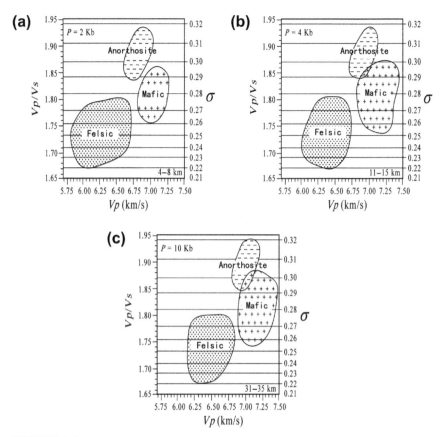

FIGURE 2.1   Poisson's ratio or V_p/V_s versus P-wave velocity under different pressures of (a) 2 kb (4- to 8-km depth), (b) 4 kb (11- to 15-km depth), and (c) 10 kb (31- to 35-km depth).

Recalling the measurement results of seismic wave velocities of the rocks from the upper mantle in recent years, we understand the following rules:

1. The types of rocks from the upper mantle can be classified into two categories. Class A is pyrolite with a low FeO content from the upper mantle, containing mainly olivine and orthopyroxene. Rocks of this class are hard to be melted. Class B is piclogite, containing mostly clinopyroxene and garnet but less of olivine and orthopyroxene. The composition of these rocks is similar to that of the basaltic magma and easy to be melted. The P-wave velocity of class A rocks is lower than that of class B rocks. The S-wave velocity of class B rocks is usually lower than that of class A rocks in the condition under a 100-km depth.

TABLE 2.2   Seismic Properties of Crystalline Rocks (Room Temperature)

Rocks	Density (g/cm^3)	V_p (km/s)	V_s (km/s)	V_p/V_s	Poisson's Ratio	Region
Spinel lherzolite		8.2	4.77		0.24	
Harzburgite		8.35–8.4	4.87–4.9		0.24	
Dunite		8.45	4.9		0.25	
Lherzolite		8.31	4.81		0.25	
Gabbro	3.03	7.43	4.537		0.20–0.26	China
Pyroxenite		7.25	4.18	1.75	0.25–0.26	Dabie
Eclogite		8.3–8.6	4.5–4.7		0.28–0.29	
Basalt	2.8	5.65	2.9		0.23–0.32	China
Basic granulite		6.65	3.82	1.78	0.25–0.27	Dabie
Diorite	2.67	5.4	3.2–3.3		0.23–0.26	China
Serpentine		6.5		1.92		Sulu
Granite	2.75	5.6–5.73	3.3–3.6		0.17–0.23	China
Gneiss		5.69	3.87	1.47–1.56	0.069–0.15	Sulu

As a result, the ratio of V_p/V_s might be used to divide these two classes of mantle rocks.

2. Velocities V_p and V_s of rocks in tectonically active regions are lower than those of craton areas. Below the crust rocks in the tectonically active regions, low-velocity and low-quality factor Q may indicate partial melting of mantle rocks.

3. In craton areas, a sudden decrease of V_p and V_s at the bottom of the lithosphere may be related to the high geothermal gradient or anisotropic effect therein. A low ratio of V_p/V_s at a depth of 50–150 km may show that the uppermost mantle is of a low geothermal gradient and rich in olivine, containing less of FeO. At depths >150 km, increasing the V_p/V_s ratio may indicate an increasing composition of garnet and clinopyroxene. The increasing FeO amount enhances the fusibility of the lower lithosphere. This rule is consistent with that of the study on the kimberlite enclosures. Compared with related observations, the oceanic mantle may contain more eclogite than the continental, so it appears less anisotropic.

The seismic properties of sedimentary rocks are discussed as follows: Unlike crystalline rocks, sedimentary rocks often contain pores or cracks, and the filling fluids have an extensive influence on physical properties. Table 2.3 shows the experimental results of the physical properties of sedimentary rocks under room temperature (Sheriff and Geldart, 1983; Mavco et al., 2009). From the table, the Poisson's ratio of water-saturated sandstone, mudstone, and shale are the highest, ranging from 0.28 to 0.34. The Poisson's ratio of limestone and anhydrite is about 0.3. The P-wave velocity of fluids decreases regularly from water to light oil, and then to gas. Poisson's ratio increases obviously in water-saturated sandstone. For the wave velocity of heavy oil, the situation is changed. The velocity of heavy oil can be higher than that of water. The Poisson's ratio of sandstone saturated with oil may decrease. For the reservoirs of sandstone and carbonate saturated with natural gas, Poisson's ratios can be <0.15, and those of tuff and dolomite can also be <0.2. Figure 2.2 shows the measurements of the V_s/V_p ratio in sandstone with the interstitial water and gas. The V_s/V_p ratio of sandstone filled with water is lower than that of sandstone filled with gas. For the sandstone filled with gas, V_p decreases and V_s does not change much, so the ratio increases.

As mentioned above, the influence of the fluid filling cracks is considerable on both wave velocity and Poisson's ratio. The fluid changes the P-wave velocity but does not change the S-wave velocity much. For example, the P-wave velocity of dry sandstone with a porosity >20% is 3.2 km/s, and the V_s/V_p ratio equals 0.75. If sandstone is saturated with water, V_p can increase to the range of 4.0–4.4 km/s, and the S-wave velocity remains stable at 2.4 km/s. The P-wave velocity of this sandstone saturated with oil will reach 4.7 km/s and the corresponding V_s/V_p ratio drops to 0.51. For carbonate reservoir rocks, wave velocities increase slightly when it is saturated with oil and water; however, the velocity increase is much smaller than that of the high-porosity sandstone, and the S-wave velocity is still not changed. Figure 2.3 shows the statistical results of the V_s/V_p ratio versus V_p for general sedimentary rocks. It can be seen that the V_s/V_p ratio of sandstone saturated with gas is almost proportional to V_p. The V_s/V_p ratio of limestone is slightly inversely proportional to V_p.

Due to velocity V_p representing bulk compressibility, the diagram of Poisson's ratio versus velocity V_p (Figure 2.4) can be used to distinguish the lithology of sedimentary rocks and saturation of fluid. The P-wave velocity of sandstone is usually <5 km/s, the Poisson's ratio of water-saturated sandstone is around 0.25. P-wave velocities of coal and mudstone shale are in the range of 3.0–4.4 km/s. P-wave velocity of carbonate is >4 and increases slightly when the carbonate is saturated with oil and water. The Poisson's ratio of carbonate is >0.3, which is higher than that of sandstone and almost the same as that of anhydrite. However, anhydrite has a higher P-wave velocity. From Figure 2.4, the Poisson's ratio of dolomite and sandstone saturated with gas is <0.2, but the V_p of dolomite is higher than that of sandstone. For delineation of

TABLE 2.3 Measurement Results of the Seismic Properties of Sedimentary Rocks (Room Temperature)

Rock	Density (g/cm³)	V_p Dry	V_p Water Saturated	V_p Saturated with Oil	V_s Dry	V_s Water Saturated	Poisson's Ratio	Region
High-porosity sandstone	2.2	3.3	4.35	4.75	2.2	2.4	Dry 0.1 Wet 0.28	Daqing
		3.2	4.0		2.0	1.95	Dry 0.18 Wet 0.34	America
Low porosity sandstone	2.4–2.7	4.5–5.2		Gas bearing 3.5–4.5	2.7–3.2	Gas bearing 2.1–2.7	Dry 0.2–0.22 Gas pores 0.12–0.22	Ordos
Siltstone	2.4						0.29	Cretaceous
Mudstone	2.0–2.6	4.0–4.7			2.0–2.6		0.28–0.33	Ordos
Shale	2.55–2.73						0.28	Triassic
Tuff	2.4–2.6						0.18	Jurassic
Limestone	2.6–2.7	4.5	4.9		2.45	2.44	Dry 0.29 Water 0.34	America
Dolomite	2.7–2.9	5.8–7.0			3.2–3.7		0.26–0.33	Ordos
		6.8			4.2	4.25	0.19	
	2.7–2.84						0.24	Triassic
Fractured breccia			4.22			2.25	0.3	Engineering test

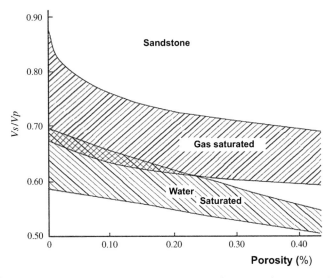

FIGURE 2.2 The measurements of velocity ratio V_p/V_s in sandstone containing pore water and gas.

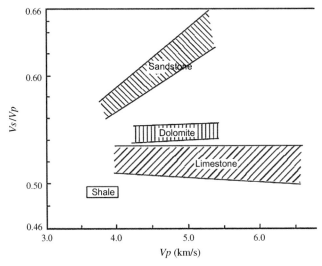

FIGURE 2.3 Measurements of velocity ratio V_p/V_s versus V_p in the main sedimentary rocks.

clastic oil and gas reservoir, it is important to determine the mud content by using Poisson's ratio. Acoustic logging has found an approximate linear relationship between the mud content r of shale and Poisson's ratio as in (Sheriff and Geldart, 1983; Mavco et al., 2009)

$$\sigma = Ar + B \tag{2.54}$$

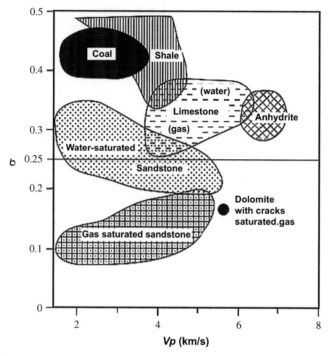

FIGURE 2.4 The measurement results of Poisson's ratio and velocity V_p of the main sedimentary rocks.

where the coefficient $A > 0$, such that $A = 0.125$ and $B = 0.18{-}0.3$. If r is in the range of $0.05{-}0.5$, Eqn (2.54) is applicable.

Quality factor Q of rocks can be obtained from seismic records on reflection explorations. The value of $1/Q$ used to be called the absorption factor for a stratum, and both Q_p and Q_s correspond to P-waves and S-waves can now be obtained from three-component records. For sedimentary strata, usually $1/Q_s > 1/Q_p$, which means the quality factors $Q_p > Q_s$. However, for most of the natural gas reservoir, Q_S^{-1} does not significantly increase, while Q_P^{-1} becomes very large. Therefore, reflective P-waves below a gas bad can be very weak, and can even completely disappear, while the event of the reflective S-wave remains very clear and continuous. Experiments show that the Q_S^{-1} of sandstone saturated with gas is lower than that of the dry rock, and when the pressure increases, Q_S^{-1} also decreases significantly. On the other hand, Q_P^{-1} is not sensitive to the variation of pressure. Experiments also show that at the critical temperature of the transition of water to vapor, the energy absorption of the reflected P-wave reaches the maximum, causing a low trough to the absorption factor Q_P^{-1}. However, the energy absorption of the S-wave caused by the phase transition does not result in a low trough of Q_S^{-1}. Therefore,

TABLE 2.4 Ultimate Strength of Common Materials (Unit: 10^7 N/m^2)

Material	Tensile Strength	Compressive Strength	Note
Cast iron	10–22	37–88	Both high
Copper	24	42	Both high
Steel	50	40	Both high
Sandstone	0.2	4–15	High compressive
Limestone		5–15	High compressive
Granite	0.3	12–26	High compressive
Pine	8.0	4.0	
Rubber	5	5	Poisson's ratio = 0.4286

the ratio of the Q_S/Q_P ratio may become an important indicator to distinguish fluid phases (Ji et al., 2002, 2007) and help detect rock strength for engineers. For strength measurements of rocks in geotechnical engineering, one needs to study the relationship between seismic velocity or Poisson's ratio and the ultimate strength of rocks. The ultimate strengths of rocks can be classified into ultimate compressive strength and the ultimate tensile strength (Table 2.4). This table shows that the compressive strength of fresh crystalline rocks is generally high, but its tensile strength is generally low. This is why normal faults have been densely developed in the upper crust.

In geotechnical engineering, only the compressive strength of rocks is usually considered, which can be calculated from seismic velocities or Poisson's ratio. By setting the compressive ultimate strength as T, shear modulus as μ, the density as ρ, Poisson's ratio as б, we can use the following empirical equation:

$$T = 0.5\left[\rho V_p^2(1 - 2\sigma)\right] \Big/ \left[C_p(1 - \sigma)\right] \qquad (2.55)$$

where the coefficient C_p is related to rock types. If the S-wave velocity is measured, the empirical equation can be rewritten as

$$T = \mu/C = V_s^2\rho/C_s \qquad (2.56)$$

where the coefficient C_S is again related to rock type. The coefficient for sandstone equals 55–180, for limestone it is 240, and for quartzite or gneiss it equals 180. Note that this empirical equation varies from place to place, and it is best to refer to local measurements and to modify the coefficients constantly.

From Elastic Waves to Seismic Waves

Acoustic waves and elastic waves based on continuum mechanics have been discussed in Chapters 1 and 2, and seismic waves will be discussed in this chapter. Seismic waves are waves propagating in the heterogeneous Earth, while acoustic waves and elastic waves in physics propagate in homogeneous media. Because seismic waves propagate in more complex media, such as the ocean bottom, rocks, and the outer core of the Earth, there are six theoretical problems that must be considered for the development of the seismic wave theory from elastic waves.

1. The forward modeling of the wave equation with variant coefficients in wave equations and the solution properties of the forward problems.
2. The coherences and differences between seismic records and ideal elastic waves.
3. The propagation of elastic waves in multilayered media that can be treated as a simplified model of sedimentary basins.
4. The propagation of elastic waves in multiphase media such as sedimentary rocks.
5. The propagation of waves in a rheological melting mass such as magma.
6. The models of the continuum mechanics for complex stratigraphic structures with different boundary layers.

Problems 1–4 will be discussed in this chapter, and question 6 will be discussed in Chapter 4.

3.1. ON ACOUSTIC WAVE EQUATIONS WITH VARIANT COEFFICIENTS

In Chapters 1 and 2, we had discussed acoustic waves and elastic waves based on continuum mechanics, and have finally deduced acoustic wave and elastic wave equations with constant coefficients. The propagation of waves in homogeneous media cannot be directly applied to the exploration of the earth, where the aim is to detect the heterogeneities of the Earth.

Mathematically (Tiknonov and Samarskii, 1963; Sobolev, 1964; Frittman, 1965), the initial and boundary value problems for acoustic wave propagation with the variant velocities $c(x)$ are much more complicated than that of the common hyperbolic differential equations such as Eqn (2.26). Because wave function $u(x)$ is related to the variant velocity, it becomes a functional $u(x,c(x))$, and new mathematical tools must be used to describe this functional. To avoid functional analysis and oversimplify the physical model as well, engineers may substitute the constitutive equation into the equation of motion with variant parameters of media as discussed in Section 1.6, and the generalized wave functions can be treated as ordinary functions and used in reflection seismology.

When the body force in the media is absent, we can obtain a set of wave equations with variant velocities as shown:

$$\frac{\partial^2 u}{\partial t^2} = \frac{\partial}{\partial x}\left[c_p^2\left(\frac{\partial u}{\partial x} + \frac{\partial v}{\partial y} + \frac{\partial w}{\partial z}\right)\right] \tag{1.50a}$$

$$\frac{\partial^2 v}{\partial t^2} = \frac{\partial}{\partial y}\left[c_p^2\left(\frac{\partial u}{\partial x} + \frac{\partial v}{\partial y} + \frac{\partial w}{\partial z}\right)\right] \tag{1.50b}$$

$$\frac{\partial^2 w}{\partial t^2} = \frac{\partial}{\partial z}\left[c_p^2\left(\frac{\partial u}{\partial x} + \frac{\partial v}{\partial y} + \frac{\partial w}{\partial z}\right)\right] \tag{1.50c}$$

The above equations are not typical acoustic equations and must be solved simultaneously due to the three-components coupling with each other. In the case that the sedimentary basin can be simplified as piled horizontal strata, the velocities of the strata can be approximated to an interval velocity function $c_p(x) = c(x,y,z)$; here $p=1,2,...$, denoting sequential number of the strada, y-axis variation of the velocities is ignored. Then, velocity $c_p(x)$ can be extracted from its average with respect to z. Under the acceptable condition in single component and near-vertical reflection method,

$$\frac{\partial w}{\partial z} \gg \frac{\partial u}{\partial x}, \quad \frac{\partial w}{\partial z} \gg \frac{\partial v}{\partial y},$$

Then, Eqn (1.50c) might be simplified as the typical acoustic wave Eqn (1.19c).

Engineers generally use the function $V(x, z)$ instead of the constant c for the acoustic wave Eqn (1.19), and obtain a 2D scalar wave equation in heterogeneous media as

$$\frac{\partial^2 u}{\partial x^2} + \frac{\partial^2 u}{\partial z^2} - \frac{1}{V(x, z)^2}\frac{\partial^2 u}{\partial t^2} = -F \tag{3.1}$$

For mathematicians, Eqn (1.19) and Eqn (3.1) are different, because Eqn (1.19) is an equation with constant coefficients and its solution of the initial and boundary value problems belongs to the category of ordinary functions, while Eqn (3.1) is an equation with variant coefficients and its solution belongs to the category of hyperfunctions (generalized functions).

Wave equations with constant coefficients have been discussed in detail in some books on Mathematical and Physical Equations (Futterman, 1962; Riley, 1974; Chen, 1979). These types of equations have some advantageous properties, such as the invariance with coordinates shifting, symmetry for the variation of time, the existence of a basic solution can be proved, the derivatives of any sufficient smoothing solution is still the solution of the original equation, and so on. However, Eqn (3.1) with variant coefficients may not have these advantageous properties. In fact, the existence of its basic solution is still uncertain.

The study on mathematical and physical equations has shown that the equations with constant coefficients and those with variant coefficients are different in the solution existence and uniqueness. The solution of the latter can only be found after their conjugate or adjoint equations have been studied. The initial and boundary value problems of the wave equations with variant coefficients can only have generalized solutions, and existence of the unique generalized solution is contingent on additional conditions. Furthermore, whether the generalized solution does or does not exist depends on the boundary type, which means that in the case of the same equations and initial conditions, the solutions for some kinds of boundaries may exist, but those of other kinds may not. The examples given by Chen Qingyi (1979) showed that there are no solutions for the hyperbolic equations when the signs of gradients change on the boundary.

Sobolev (1963), a mathematician of the former Soviet Union, studied the vibration of a string with different densities on the two sides. He found that there were no classic solutions for the wave equations with variant coefficients, which means that the solutions cannot be expressed by any analytical functions. But it is possible to find reasonable generalized solutions based on the variation principles (Riley, 1974; Aubin, 1979; Hutson and Pym, 1980; Mrinal and Stoffa, 1995), such as the principle of minimum energy. He pointed out that "In order to make more wave equations with variant coefficients solvable, one has to study the generalized functions as much as possible. However, the uniqueness for the generalized solutions should be promised." He suggested researching the generalized solutions from the family of square integral functions, but not from the family of square differentiable functions. For the wave field vector u, one may look at minimum of the functional

$$J(u) = \iiint \left(u_x^2 + u_y^2 + u_z^2 \right) dx\, dy\, dz$$

The uniqueness of the solution of minimum $J(u)$ can be proved by the uniqueness theorem.

Solving the wave equations with variant coefficients combining with discontinuous boundary values may give rise to some new questions. The discontinuous boundary values may be generated since the parameters of heterogeneous media are variable in the adjacent areas of the boundary. In this case, the solutions of the equations with the initial value and boundary value are usually not unique, that is no classic analytical solutions can be found. It is necessary to add some additional conditions to ensure that the solutions of hyperbolic equations for a given initial value and boundary values can become unique (Marozov, 1984). However, the additional conditions proposed by mathematicians might not be physically reasonable. For example, the initial value should be an analytical function according to Cauchy's theorem. However, the impulse function for shots in reflection seismology is not the typical analytical function.

To compensate for the limitations on the physical models mentioned above, the term "boundary layer" and the conditions for the boundary layer are introduced in continuum mechanics, which is called the "boundary layer theory"(Fung, 1977; Spencer, 1980; Du Xun, 1985). The boundary layer is just a simplification of the real interface between two sedimentary layers and represents the boundary within the continuous media, in which the boundary is represented as a complex thin layer. The boundary layer condition is adaptable for studying the entire wave field outside of the thin boundary layer, but is not adaptable for studying the wave field inside of the thin layer that can be ignored in reflection seismology. The application of continuum mechanics is related to the scale of objects under study and is not applicable to deal with small-scale inhomogeneous objects very close to the boundary. Mathematicians have found that the nonuniqueness of solutions of the hyperbolic equations with the initial and boundary values will occur in some small local areas, that is around some singularity points, where the microscale wave field cannot be described accurately with continuum mechanics. In classic reflection seismology, the boundary layer theory is acquiesced. For example, only the reflection wave and transmission wave on the boundary are usually taken into account, and not the grain motion being investigated. However, if a rock layer becomes very thin, much thinner than half of the seismic wavelength, then the boundary layer theory may not be so suitable for wave field discretion.

In the discussion above, we use classic mathematical tools to expand the mathematical methods from the wave equations with constant coefficients to that with varying coefficients in a certain limited range. We must answer the questions, "What are the properties of the solutions of the wave equations with variant coefficients"? Are there any standard mathematical methods for solving these wave equations with the initial and boundary value? The solutions of the wave equation's initial and boundary value problems with variant coefficients depend on the properties of its partial differential operators; therefore, this study must rely on the theory of operators. Since the middle of the twentieth century, scientists have developed new mathematical tools for studying the partial differential equations with variable coefficients, which are known as Fourier integral operators and pseudodifferential operator. Both the two theories are too abstruse to be explained clearly in this section, and will be discussed further in Sections 6.5 and 6.6. Before the discussion, we will still expand the wave equations with constant coefficients to the ones with variant coefficients approximately, and take the generalized wave functions as ordinary wave functions. In Section 6.5, we will prove that a wave operator with constant coefficients is the first-order approximation of the wave operator with variant coefficients.

For the seismic wave propagated in heterogeneous media, the common strategy for simplifying Earth models is by constraining the media between the interfaces and treating the interfaces with boundary conditions. The velocity between two adjacent interfaces can be reduced as a constant or the interval velocity. Therefore, wave operators with constant coefficients can be used for describing the propagation of seismic waves, and soft boundaries are inserted into the interfaces by ignoring the details of seismic waves inside the boundaries. This strategy is in accordance with that of continuum mechanics, and adapted by most seismic scientists. The layered earth model is the most popular model in reflection seismology and can be used to study the propagation of seismic waves in horizontally layered earth (Section 3.3).

For a heterogeneous model of layers that are not horizontal and have laterally variant velocities, wave equations with variant coefficients can be used in the following presumptions: (1) the interfaces are defined by a continuous monodromy function $z_i(x,y)$, where $i = 1, 2,...$. (2) The elastic parameters between interfaces, such as $V_{pi}(x,y,z)$, must be smooth functions. In fact, it is common that $V_{pi}(x,y,z)$ is replaced by the internal velocity $V_{pi}(x,y)$ by using the vertical average velocity within a layer. (3) If there are no terms of the first-ordered derivative of the wave field with respect to t in wave equations, that is one ignores the wave dispersion and attenuation, the generalized solution of the acoustic equation with variant coefficients is existent. Thus, the generalized solution can be used to approximate the solution functions for the layered earth models.

Many kinds of artificial seismic sources are used in reflection seismology (Sheriff, 1983, 1984; Telford, 1990). For example, we use explosion or vibrators on the land and use high-pressure air guns, water guns, electrical sparks as sources in marine reflection. Sources are called shots or shooting in seismic exploration and produce the body force term on the right-hand side of wave equations. The usual mathematic models of sources are the point body force. For oil/gas exploration, the energy of sources is not necessarily very strong, and the point body force is sufficient to describe the sources.

Mathematically, the δ function (Frittman, 1965), which represents a point source, denotes the location of the source. The function $W(t)$ denotes the source amplitude and waveform, called the seismic wavelet and having different kinds. For example, the wavelet generated by explosive sources can be presented by the Ricker wavelet, the one generated by vibrators can be presented by the Cloutier wavelet, and so on. Assuming that the initial value is $u|_{t=0} = 0$, the scalar acoustic wave equations in the isotropic media can be rewritten as follows:

$$\frac{\partial^2 u}{\partial t^2} = W(t)\delta(\mathbf{x} - \mathbf{x}_s) + c^2 \left[\frac{\partial^2 u}{\partial x^2} + \frac{\partial^2 u}{\partial y^2} + \frac{\partial^2 u}{\partial z^2}\right] \tag{3.2}$$

where $x = (x,y,z)$ is any point and x_s is the location of the point source. The default initial condition of the equation is

$$u(\mathbf{x}, t) = 0 \quad \text{for} \quad t \leq 0$$

The general solution in polar coordinates can be obtained when c is a constant (see more details in Section 5.1):

$$u(\mathbf{x}, t) = \frac{1}{4\pi c^2} \frac{W\left(t - \frac{|x - x_s|}{c}\right)}{|x - x_s|} \tag{3.3}$$

The above equation shows that, under the condition of the point source, the wavelet function $W(t)$ shifts with increasing time with no change in the waveform, and the amplitude decreases with increasing distance away from the source, which is called "the geometric decay". Of course, this method is just an approximate description of the acoustic far wave field in homogeneous media.

When the receivers are very close to the source and the source is relatively big enough, the source cannot be treated as a point source anymore but as a block force whose volume is denoted as V. The block force function can be expressed by a function $W(x,t)$ of spatial variants and time; then, the wave equation can be written as follows:

$$\frac{\partial^2 u}{\partial t^2} = \frac{W(\mathbf{x}, t)}{\rho} + c^2 \left[\frac{\partial^2 u}{\partial x^2} + \frac{\partial^2 u}{\partial y^2} + \frac{\partial^2 u}{\partial z^2}\right] \tag{3.4}$$

Since the block body force can be considered as an assembly of the point force, according to the superposition principle, the general solution of above equation becomes

$$u(\mathbf{x}, t) = \frac{1}{4\pi\rho c^2} \iiint \frac{W\left(\mathbf{x}_s, t - \frac{|x - x_s|}{c}\right)}{|x - x_s|} d\mathbf{x}_s, \quad x_s \in V \tag{3.5}$$

Equation (3.5) gives the basic solution of the acoustic wave equation in homogeneous media, and it is more accurate than Eqn (3.3) for a near-source wave field. However, the wavelet function does not express the real force, and the acoustic equation is used rather than the elastic equations. Therefore, solution Eqn (3.5) is not the accurate solution. Aki and Richards (1980) discussed the isotropic elastic wave equations for the near-source wave field and proved that Eqn (3.5) is consistent with the solution of the far wave field in homogeneous media. The solution of the near-source wave field is more complex, including the high-order attenuation term of $|x - x_s|$. In this book, by default, the wave equations are expressed mathematically with a point source or a plane wave source.

3.2. SEISMIC REFLECTION RECORDS AND CORRESPONDING EQUATIONS

Geophysicists work on seismic records that used to be called seismograms and contain recorded seismic wave fields. We should explain the relationship of seismic records and their corresponding elastic wave equations.

3.2.1. Wave Equations for Marine Reflection Records

In marine seismic exploration, both the point sources and receivers are located under water, and the seawater is a natural filter for shear waves. The point sources generate pressure waves and receivers receive the scalar compressive wave field (Sheriff and Geldart, 1983; Telford et al., 1990). Although there are secondary shear waves in rocks under the seabed, it can be treated as noise because the recorded wave field is filtered by seawater, and has only weak noises in records. So marine seismograms can be expressed using scalar acoustic wave Eqns (3.1) and (3.2) as

$$\left[\frac{\partial^2 u}{\partial x^2} + \frac{\partial^2 u}{\partial y^2} + \frac{\partial^2 u}{\partial z^2}\right] - \frac{1}{c^2(x,y,z)}\frac{\partial^2 u}{\partial t^2} = -W(t)\delta(x_s, y_s) \qquad (3.6)$$

where $W(t)$ is the wavelet of the point source, (x_s, y_s) are the coordinates of the point source. The initial condition is $u(x,y,z,t) = 0$ when $t < 0$. Here we assume that the source depth $z_s = 0$ after the de-ghost filtering of seismic records (see section 4.2 for details).

In the following expressions on the boundary conditions, the square brackets "[*]" denote the differences of stress or strain on the two sides of an interface. For the air/seawater interface, only the stress and vertical components of the strain are continuous:

$$[A_{zz}] = 0, \quad [u] \approx 0 \qquad (3.7)$$

For the seawater/rock interface, the horizontal components of strain on the boundary is not continuous, while the stress and vertical strain are continuous:

$$[A_{zx}] = 0, \quad [A_{zy}] = 0, \quad [A_{zz}] = 0; \quad [u] \approx 0 \qquad (3.8)$$

Theoretically, the more accurate method to describe a marine seismogram is to use elastic wave equations to describe the wave field under the seabed and assume the shear modulus of seawater as zero. The problem is how to convert the component of the displacement from the elastic equation into measured acoustic pressure. In recent years, three-component geophones are set on the seabed for recording seismograms in marine exploration. The use of the elastic wave equations might become indispensable to describe the wave field under the seabed.

3.2.2. Wave Equations for Land Single-Component Records

Land 2D seismic survey records of the vertical component are obtained with both the point source and receivers located in soils on the earth's surface. Since the soil layers usually are of a low velocity, the energy of the vertical component of upgoing reflection waves is dominant in land records, through relatively weak shear wave exits. Single-component receivers observe and record the vertical component of displacement wave fields $w(x,y,z,t)$. Since the earth's surface is not homogeneous, terrain changes and geophones imperfect coupling with soil, the S/N ratio of the seismic data for land surveys is usually lower than that in marine reflection surveys. Thus, the single-component land 2D seismogram can be approximately described by the scalar elastic wave Eqn (2.26c) as follows:

$$(\lambda + 2\mu)\frac{\partial^2 w}{\partial z^2} + \mu\left(\frac{\partial^2 w}{\partial x^2} + \frac{\partial^2 w}{\partial y^2}\right) + (\lambda + \mu)\left(\frac{\partial^2 v}{\partial y \partial z} + \frac{\partial^2 u}{\partial x \partial z}\right) - \rho\frac{\partial^2 w}{\partial t^2}$$
$$= -W(t)\delta(x_s, y_s, z_s) \tag{3.9}$$

where $W(t)$ is the wavelet function of a point source and (x_s, y_s, z_s) is the location of the point source. The 2D seismogram can be obtained from the derivative of Eqn (3.9) with respect to y and setting it to zero as follows:

$$(\lambda + 2\mu)\frac{\partial^2 w}{\partial z^2} + \mu\frac{\partial^2 w}{\partial x^2} + (\lambda + \mu)\frac{\partial^2 u}{\partial x \partial z} - \rho\frac{\partial^2 w}{\partial t^2} = -W(t)\delta(x_s, z_s) \tag{3.10}$$

Since the horizontal component u and v, and vertical component w are coupled with each other, the single-component seismogram on land could be converted into the wave field of the acoustic equation only after a special procedure of data processing, which will be discussed in the next chapter.

3.2.3. Wave Equations for Land Three-Component Records

The three-component seismic data recorded by three-component geophones on land are written as (u, v, w), where w usually has a higher S/N ratio than the others have. Three-component seismograms on land can be described approximately by the elastic wave Eqn (2.26), according to the strategy of generalizing the equation with constant coefficients to that with variant coefficients and treating the generalized wave functions as ordinary functions. Thus, we have

$$(\lambda + 2\mu)\frac{\partial^2 u}{\partial z^2} + \mu\left(\frac{\partial^2 u}{\partial x^2} + \frac{\partial^2 u}{\partial y^2}\right) + (\lambda + \mu)\left(\frac{\partial^2 v}{\partial y \partial x} + \frac{\partial^2 w}{\partial x \partial z}\right) - \rho\frac{\partial^2 u}{\partial t^2}$$

$$= -W(t)\delta(x_s, y_s, z_s)$$

$$(\lambda + 2\mu)\frac{\partial^2 v}{\partial z^2} + \mu\left(\frac{\partial^2 v}{\partial x^2} + \frac{\partial^2 v}{\partial y^2}\right) + (\lambda + \mu)\left(\frac{\partial^2 w}{\partial y \partial z} + \frac{\partial^2 u}{\partial y \partial x}\right) - \rho\frac{\partial^2 v}{\partial t^2}$$

$$= -W(t)\delta(x_s, y_s, z_s)$$

$$(\lambda + 2\mu)\frac{\partial^2 w}{\partial z^2} + \mu\left(\frac{\partial^2 w}{\partial x^2} + \frac{\partial^2 w}{\partial y^2}\right) + (\lambda + \mu)\left(\frac{\partial^2 v}{\partial y \partial z} + \frac{\partial^2 u}{\partial x \partial z}\right) - \rho\frac{\partial^2 w}{\partial t^2}$$

$$= -W(t)\delta(x_s, y_s, z_s) \tag{3.11}$$

For air/rock interfaces, the horizontal components of the displacements on the boundary are discontinuous, while both the stresses and vertical displacement component on the boundary are continuous (Eringen and Suhubi, 1975):

$$[A_{zx}] = 0, \quad [A_{zy}] = 0, \quad [A_{zz}] = 0; \quad [w] \approx 0.$$

For the rock/rock interfaces, both the displacements and stress components are all continuous on the boundary

$$[A_{zx}] = 0, \quad [A_{zy}] = 0, \quad [A_{zz}] = 0; \quad [u] \approx 0, \quad [v] \approx 0, \quad [w] \approx 0.$$

In the vertical seismic profiling (VSP), receivers are located in the holes, far away from the ground. The air/rock interface can be ignored in VSP and crosshole surveys. It is necessary to distinguish "soft boundary" and "hard boundary" because the boundary conditions usually generate different seismic phases, acting similar to reflectors. The so-called hard boundary is the boundary where the reflection energy is strong compared to the transmission energy, such as the air/rock boundary. This type of boundary is generally used as the outermost boundary of study domains, and the wave field outside the boundary can be ignored mathematically. The so-called soft boundary is the boundary where the reflection energy is far weaker than the transmission energy, such as the interfaces between sedimentary rock layers. We have to include this type of boundaries in our study, considering the wave field effect outside the boundaries. The seabed boundary at which the reflection energy and transmission energy are nearly the same can be called the relatively hard boundary. Since the hard or relatively hard boundaries can generate strong multiples and converted waves, it is difficult to remove these boundary conditions to the source term in a wave equation. However, the soft boundaries do not excite strong multiples and can sometimes be removed to the source term of the equations together with some given boundary values. This is why we want to distinguish the "soft" boundaries from the "hard" ones. Although this distinction is not important in mathematics and physics, it is valuable in reflection seismology, because the investigation targets of seismic exploration are just "soft boundaries".

As mentioned before, it may not be accurate to generalize the wave Eqn (3.11) with constant coefficients to that with variant coefficients directly. We can improve the results a little bit by resuming deduction with the scalar wave equations that are corresponding to the three displacement components of the elastic wave equation with variant coefficients. In 3D linear isotropic elastic media, by substituting stress A and strain E in Eqn (2.9c) into Hooke's law, we can obtain the following set of scalar constitutive equations:

$$
\begin{aligned}
A_{xx} &= (\lambda + 2\mu)E_{xx} + \lambda E_{yy} + \lambda E_{zz} \\
A_{yy} &= \lambda E_{xx} + (\lambda + 2\mu)E_{yy} + \lambda E_{zz} \\
A_{zz} &= \lambda E_{xx} + \lambda E_{yy} + (\lambda + 2\mu)E_{zz} \\
A_{xy} &= \mu E_{xy}, \quad A_{yz} = \mu E_{yz}, \quad A_{zx} = \mu E_{zx}
\end{aligned}
\tag{3.12}
$$

where the Lame coefficients $\lambda(x)$ and $\mu(x)$ are continuous functions of spatial coordinates $x = (x,y,z)$. According to stress equilibrium Eqn (2.2), the summation of the divergences of the stresses on the six faces of a mass grain is almost equal to the volume force. Substituting Eqn (2.2) into the equation of motion with the absence of body force yields

$$
\rho \frac{\partial^2 \mathbf{u}}{\partial t^2} = div\mathbf{A}
\tag{2.3}
$$

Assuming that the density is a constant, the P-wave velocity is $c_p(x) = \sqrt{\frac{\lambda + 2\mu}{\rho}}$, and the velocity of the shear wave is $c_s(x) = \sqrt{\frac{\mu}{\rho}}$; substituting Eqn (2.6) and Eqn (3.12) into Eqn (2.3) and then expanding it, we obtain

$$
\begin{aligned}
\frac{\partial^2 u}{\partial t^2} &= \frac{\partial}{\partial x}\left[c_p^2\left(\frac{\partial u}{\partial x} + \frac{\partial v}{\partial y} + \frac{\partial w}{\partial z}\right)\right] - 2\frac{\partial}{\partial x}\left[c_s^2\left(\frac{\partial v}{\partial y} + \frac{\partial w}{\partial z}\right)\right] \\
&\quad + \frac{\partial}{\partial y}\left[c_s^2\left(\frac{\partial u}{\partial y} + \frac{\partial v}{\partial x}\right)\right] + \frac{\partial}{\partial z}\left[c_s^2\left(\frac{\partial w}{\partial x} + \frac{\partial u}{\partial z}\right)\right]
\end{aligned}
\tag{3.13a}
$$

$$
\begin{aligned}
\frac{\partial^2 v}{\partial t^2} &= \frac{\partial}{\partial y}\left[c_p^2\left(\frac{\partial u}{\partial x} + \frac{\partial v}{\partial y} + \frac{\partial w}{\partial z}\right)\right] - 2\frac{\partial}{\partial y}\left[c_s^2\left(\frac{\partial u}{\partial x} + \frac{\partial w}{\partial z}\right)\right] \\
&\quad + \frac{\partial}{\partial z}\left[c_s^2\left(\frac{\partial v}{\partial z} + \frac{\partial w}{\partial y}\right)\right] + \frac{\partial}{\partial x}\left[c_s^2\left(\frac{\partial u}{\partial y} + \frac{\partial v}{\partial x}\right)\right]
\end{aligned}
\tag{3.13b}
$$

$$\frac{\partial^2 w}{\partial t^2} = \frac{\partial}{\partial z}\left[c_p^2\left(\frac{\partial u}{\partial x} + \frac{\partial v}{\partial y} + \frac{\partial w}{\partial z}\right)\right] - 2\frac{\partial}{\partial z}\left[c_s^2\left(\frac{\partial u}{\partial x} + \frac{\partial v}{\partial y}\right)\right]$$
$$+ \frac{\partial}{\partial x}\left[c_s^2\left(\frac{\partial w}{\partial x} + \frac{\partial u}{\partial z}\right)\right] + \frac{\partial}{\partial y}\left[c_s^2\left(\frac{\partial v}{\partial z} + \frac{\partial w}{\partial y}\right)\right]$$

(3.13c)

The set of Eqn (3.13) represents the elastic wave field corresponding to three-component 3D land seismic records. As variant velocity has been included, they can be used after improvement for reflection seismic exploration on land, VSP, and crosshole seismic tomography.

3.3. ELASTIC WAVES IN HORIZONTALLY MULTILAYERED MEDIA

For the seismic waves propagating in heterogeneous media in the Earth, we can simplify the Earth heterogeneity models by the layered earth model. This model uses many interfaces and treats these interfaces as soft boundaries, assuming that the velocities between interfaces are constants. Therefore, the wave operators with constant coefficients can be used to express the propagation of seismic waves in the heterogeneities in the earth. In this section, we will use the strategy to study the propagation of seismic waves in horizontally layered media. The aim of seismic exploration is to find oil and gas, which costs about one-third of the funding involved in oil and gas exploration. As the exploration is usually carried out in sedimentary basins, the physical model of sedimentary basins is often used in the study of reflection seismic waves. As the simplest model, the propagation of seismic waves in horizontal isotropic multilayered media is the basic topic in reflection seismology. In the twentieth century, many scientists have been attracted by this topic and have made great progress in it (Ewing et al., 1957; Kennett, 1981). Kennett not only derived propagation matrices and revealed the quantitative relationship between different seismic wave phases but he also showed the analytical expressions of the three decoupling components of the displacement vector. In this section, we will follow Kennett's derivation to show his main results.

3.3.1. Elastic Wave Equations in a Cylindrical Coordinate System

Since the horizontal multilayered isotropic elastic layers are homogeneous within a layer and the interfaces are boundaries between adjacent layers are soft boundaries, the special parameters can be defined by functions with respect to depth z. If the point source is located on the Z axis, it is best to study elastic wave problems in cylindrical coordinates.

A point in cylindrical coordinates is expressed by (r, φ, z). We denote $\rho(z)$ as the density, $u(r, \varphi, z, t)$ as the displacement vector, and (u_r, u_φ, u_z) as its three components. In constitutive Eqn (2.12), all the stress tensors and strain tensors in homogeneous media will be rewritten as their corresponding forms in cylindrical coordinates. Here, $f = (f_r, f_\varphi, f_z)^T$ denotes the body force or the source, and the stress tensor and strain tensor are written as follows:

$$A = \begin{bmatrix} A_{rr} & A_{r\varphi} & A_{rz} \\ A_{r\varphi} & A_{\varphi\varphi} & A_{\varphi z} \\ A_{rz} & A_{\varphi z} & A_{zz} \end{bmatrix}; \quad E = \begin{bmatrix} E_{rr} & E_{r\varphi} & E_{rz} \\ E_{r\varphi} & E_{\varphi\varphi} & E_{\varphi z} \\ E_{rz} & E_{\varphi z} & E_{zz} \end{bmatrix}$$

According to the rule of coordinate transformation in analytical geometry, the equation of motion, which the stress tensor follows, can be written as

$$\partial_z A_{rz} + \partial_r A_{rr} + r^{-1}\partial_\varphi A_{r\varphi} + r^{-1}\left(A_{rr} - A_{\varphi\varphi}\right) = \rho\left(\partial_{tt} u_r - f_r\right) \tag{3.14a}$$

$$\partial_z A_{\varphi z} + \partial_r A_{r\varphi} + r^{-1}\partial_\varphi A_{\varphi\varphi} + 2r^{-1}\left(A_{r\varphi}\right) = \rho\left(\partial_{tt} u_\varphi - f_\varphi\right) \tag{3.14b}$$

$$\partial_z A_{zz} + \partial_r A_{rz} + r^{-1}\partial_\varphi A_{\varphi z} + r^{-1}(A_{rz}) = \rho\left(\partial_{tt} u_z - f_z\right) \tag{3.14c}$$

The left-hand side of Eqn (3.14) is the divergence of stress, equivalent to the surface force, and the corresponding strain tensor becomes

$$E_{rr} = \partial_r u_r, \quad E_{\varphi\varphi} = r^{-1}\left(\partial_\varphi u_\varphi + u_r\right) \quad E_{zz} = \partial_z u_z,$$
$$E_{\varphi z} = \tfrac{1}{2}\left(r^{-1}\partial_\varphi u_z + \partial_z u_\varphi\right)$$
$$E_{rz} = \tfrac{1}{2}\left(\partial_z u_r + \partial_r u_z\right) \tag{3.15}$$
$$E_{r\varphi} = \tfrac{1}{2}\left(\partial_r u_\varphi + r^{-1}\partial_\varphi u_r - r^{-1}u_\varphi\right)$$

Our aim is to deduce the analytical expressions of the three components of the displacement vector (u_r, u_φ, u_z). Kennett has pointed out that it is necessary to add some assumptions to obtain the analytical solution of the equations. The key point is to assume that the SV wave and SH wave are not coupled with each other. Furthermore, we have to find a new kind of transformation that can express the SV wave and the SH wave separately by using the three components (u_r, u_φ, u_z) in cylindrical coordinates. The displacement of the SV wave and the corresponding stress components can be expressed with this kind of transformation as

$$u_v = r^{-1}\left[\partial_r(r u_r) + \partial_\varphi u_\varphi\right] \tag{3.16}$$

$$A_{vz} = r^{-1}\left[\partial_r(r A_{rz}) + \partial_\varphi A_{\varphi z}\right]$$

Similarly, the displacement of the *SH* wave and the corresponding stress components can be expressed as

$$u_h = r^{-1}\left[\partial_r\left(ru_\varphi\right) - \partial_\varphi u_r\right]$$
$$A_{hz} = r^{-1}\left[\partial_r\left(rA_{\varphi z}\right) - \partial_\varphi A_{rz}\right] \tag{3.17}$$

The body force $F = (f_z, f_v, f_h)$, having three components, and the two horizontal components of the body force (f_v, f_h) can be obtained by using the same transformation, while the vertical component f_z does not need transformation. The Laplace operator along the direction (r, φ) in the cylindrical coordinate is expressed as

$$\Delta u = r^{-1}\partial_r(r\partial_r u) + r^{-2}\partial_{\varphi\varphi}u \tag{3.18}$$

Note that Eqns (3.16)–(3.18) do not include a derivative with respect to z, meaning that inside a layer rocks are homogeneous and the derivative of the shear wave displacement and stress with respect to z equals to zero.

Substituting Eqns (3.15)–(3.18) into the equation of motion Eqn (3.14) yields the three decoupled displacement components as

$$\partial_z u_z = -\lambda(\lambda + 2\mu)^{-1}u_v + (\lambda + 2\mu)^{-1}A_{zz}$$
$$\partial_z u_v = -\Delta u_z + \mu^{-1}A_{vz} \tag{3.19}$$
$$\partial_z u_h = \mu^{-1}A_{hz}$$

Here, we have expressions for the vertical displacement component, the displacement of *SV* wave, and that of the *SH* wave. The stress components are as follows:

$$\partial_z A_{zz} = \rho\partial_{tt}u_z - A_{vz} - \rho f_z$$
$$\partial_z A_{vz} = [\rho\partial_{tt} - \gamma\Delta]u_v - \lambda(\lambda + 2\mu)^{-1}\Delta A_{zz} - \rho f_v \tag{3.20}$$
$$\partial_z A_{hz} = (\rho\partial_{tt} - \mu\Delta)u_h - \rho f_h$$

where

$$\gamma = 4\mu(\lambda + \mu)/(\lambda + 2\mu)$$

The displacement vectors of *SV* and *SH* in Eqn (3.19) are decoupled obviously, and their velocity equals to $\sqrt{\frac{\mu}{\rho}}$. There are Laplace operators in Eqns (3.19) and (3.20), and we need to expand these operators to derive the analytic expressions of the three components of the displacement vector. To expand the Laplace operator, the Fourier–Bessel transformation should be used for the displacement solutions in the frequency domain.

With the Fourier–Bessel transformation, the coordinate r is converted to the wave number k, the coordinate φ converted to wave number m, and time t is converted to frequency ω. The vertical coordinate z does not need transformation. Therefore, with the Fourier–Bessel transformation

$$u(k, m, \omega) = (2\pi)^{-1} \int_{-\infty}^{\infty} dt\, e^{i\omega t} \int_{0}^{\infty} dr\, r J_m(kr) \int_{-\pi}^{\pi} d\varphi\, e^{-im\varphi} u(r, \varphi, t) \quad (3.21)$$

the displacement $u(r, \varphi, z, t)$ is converted into $u(k,m,z,\omega)$. Similarly, the same procedure can be used for stress transformation. The transformation (3.21) yields

$$\Delta u(k, m, \omega) = -k^2 u(k, m, \omega) \quad (3.22)$$

Equation (3.22) shows that the Fourier–Bessel transformation can convert the Laplace operator into the corresponding algebraic operation in the wave number–frequency domain. As a result, solving the differential equation is converted to solve its corresponding algebraic equation. Finally, the displacement $u(r, \varphi, z, t)$ can be obtained from $u(k,m,z,\omega)$ by the inverse Fourier–Bessel transformation.

For simplicity, set $p = k/\omega$; p indicates the slowness of the P-wave. We may use a new set of variables or functions to express displacements and stresses as

$$U = u_z(k, m, z, \omega); \quad V = -k^{-1} u_v(k, m, z, \omega); \quad W = -k^{-1} u_h(k, m, z, \omega) \quad (3.23)$$

$$P = A_{zz}(k, m, z, \omega); \quad S = -k^{-1} A_{vz}(k, m, z, \omega); \quad T = -k^{-1} A_{hz}(k, m, z, \omega) \quad (3.24)$$

where U, V, and W are the displacement wave fields of the P-wave, SV wave, and SH wave, respectively. P, S, and T are the vertical and two horizontal orthogonal shear stresses, respectively. With this strategy, the relationship between displacements and stresses can be expressed as in the following differential equations:

$$\frac{\partial}{\partial z} \begin{pmatrix} U \\ V \\ \omega^{-1}P \\ \omega^{-1}S \end{pmatrix} = \omega \begin{pmatrix} 0 & p(1 - 2\beta^2/\alpha^2) & (\rho\alpha^2)^{-1} & 0 \\ -p & 0 & 0 & (\rho\beta^2)^{-1} \\ -p & 0 & 0 & p \\ 0 & p\left[4\left(1 - \dfrac{\beta^2}{\alpha^2}\right)\beta^2 p^2 - 1\right] & -p(1 - 2\beta^2/\alpha^2) & 0 \end{pmatrix}$$

$$\times \begin{pmatrix} U \\ V \\ \omega^{-1}P \\ \omega^{-1}S \end{pmatrix} + \rho\omega^{-1} \begin{pmatrix} 0 \\ 0 \\ -f_z \\ k^{-1}f_v \end{pmatrix}$$

$$(3.25)$$

The structure of the above equation is "the derivatives of the displacement–stress vector with respect to z equals to the product of the propagation connection matrix A and the displacement–stress vectors, plus the body force f'. Similarly, for the SH wave, we have

$$\frac{\partial}{\partial z}\begin{pmatrix} W \\ \omega^{-1}T \end{pmatrix} = \begin{pmatrix} 0 & (\rho\beta^2)^{-1} \\ \rho[\beta^2 p^2 - 1] & 0 \end{pmatrix}\begin{pmatrix} W \\ \omega^{-1}T \end{pmatrix} + \rho\omega^{-1}\begin{pmatrix} 0 \\ k^{-1}f_h \end{pmatrix}$$

(3.26)

where α and β denote the velocities of the P-wave and S-wave, respectively.

Since the horizontal layered velocity model can be defined by functions of variable z, for the point source located on the z axis, set the displacement–stress vector in matrix equation (3.25) as

$$\boldsymbol{B}(k, m, z, \omega) = \left[U, V, \omega^{-1}P, \omega^{-1}S\right]^T$$

(3.27)

and denote A as the propagation connection matrices; the matrix equation of displacement–stress vectors can be written as

$$\partial_z \boldsymbol{B}(k, m, z, \omega) = \omega A(p, z)\boldsymbol{B}(k, m, z, \omega) + \rho\omega^{-1}\boldsymbol{F}(k, m, z, \omega)$$

(3.28)

Obviously, after the Fourier–Bessel transformation, the equation of motion of the displacement–stress vector is converted to a linear partial differential equation in cylindrical coordinates, and the propagation of the wave field can be expressed by using only one matrix A. Therefore, the study of the wave equation in horizontal layered media can be simplified to construct the propagation connection matrices, which will be discussed later.

Of course, we are concerned about the forward modeling of the seismogram. As the displacement $u(r, \varphi, z, t)$ could be easily obtained from $u(k, m, z, \omega)$ by inverse Fourier–Bessel transformation, we have

$$u_z = (2\pi)^{-1}\int_{-\infty}^{\infty}d\omega\, e^{-i\omega t}\int_0^{\infty}dk\, k\sum_{m=-\infty}^{\infty}U(k, m, z, \omega)J_m(kr)e^{im\varphi}$$

(3.29)

$$u_r = (2\pi)^{-1}\int_{-\infty}^{\infty}d\omega\, e^{-i\omega t}\int_0^{\infty}dk\, k\sum_{m=-\infty}^{\infty}\left[V(k, m, z, \omega)\left(\frac{\partial J_m(kr)}{\partial(kr)}\right)\right.$$
$$\left. + W(k, m, z, \omega)\frac{im}{kr}J_m(kr)\right]e^{im\varphi}$$

(3.30)

$$u_\phi = (2\pi)^{-1}\int_{-\infty}^{\infty}d\omega\, e^{-i\omega t}\int_0^{\infty}dk\, k\sum_{m=-\infty}^{\infty}\left[V(k, m, z, \omega)\frac{im}{kr}J_m(kr)\right.$$
$$\left. - W(k, m, z, \omega)\left(\frac{\partial J_m(kr)}{\partial(kr)}\right)\right]e^{im\varphi}$$

(3.31)

These expressions can be used for computing the elastic wave field generated by a point source under the assumption of $|m| < 2$. In Cartesian coordinates, Eqns (3.23) and (3.24), which express the symmetrical displacement–stress variables in cylindrical coordinates, can be rewritten as

$$U = iu_z(k, m, z, \omega); \quad V = u_x(k, m, z, \omega); \quad W = u_y(k, m, z, \omega)$$
$$P = iA_{zz}(k, m, z, \omega); \quad S = A_{xz}(k, m, z, \omega); \quad T = A_{yz}(k, m, z, \omega) \qquad (3.32)$$

Note that the wave numbers are related to the velocities α and β, and the Fourier–Bessel transformation can be used only when the velocity functions are continuous.

3.3.2. Boundary Conditions

We have to put some boundary conditions to solve the elastic wave equations in the cylindrical coordinate system. If we set the vertical stress to zero on the ground surface, the displacement–stress vector of the P–SV wave in the wave number–frequency domain becomes

$$B_p(k, m, 0, \omega) = [U, V, 0, 0]^T \qquad (3.33a)$$

The displacement–stress vector of the P–SH wave in the wave number–frequency domain is

$$B_h(k, m, 0, \omega) = [W, 0]^T \qquad (3.33b)$$

In the zone where the body force is zero, Eqn (3.28) can be written as

$$\partial_z B(k, m, z, \omega) = \omega A(p, z)B(k, m, z, \omega) \qquad (3.34)$$

For the given variables k, m, and ω, $B(z)$ should be the solution of the linear ordinary differential equation, and it satisfies

$$\partial_z B(z) = \omega A(z)B(z) \qquad (3.35)$$

where A indicates the propagation connection matrix; the corresponding real propagation matrix P should satisfy the linear algebra equation as

$$B(z) = P(z, z_0)B(z_0) \qquad (3.36)$$

Showing waves propagate from depth z_0 to depth z, and on the surface

$$P(z_0, z_0) = I \qquad (3.37)$$

I indicates the unit matrix. The relationship between the propagation connection matrix A and the propagation matrix is

$$P(z_n, z_0) = \prod_{j=1}^{n} \exp\left[\omega A(q_j)(z_j - z_{j-1})\right] \qquad (3.38)$$

where j is the sequential number of the layers, z_j is the bottom depth of the jth layer, q_j is the depth of an arbitrary point in the jth layer, and $z_{j-1} < q_j < z_j$. Equations (3.36)–(3.38) show that the propagation of elastic waves in horizontal layered media follows the rule of linear recursion. Since the converted waves are generated repeatedly on the interfaces, displacement and stress can split in the form of a cumulative product.

On the model of two horizontal layers, Kennett (1983) has discussed in detail on how to construct propagation matrices. Theoretically, his results are more accurate than the equation solutions by assuming the incident planar waves, because a point source is included. The corresponding wave equations with assuming the incident planar waves can be found in many books on seismic exploration; we will discuss only its application in reflection seismic modeling as follows.

3.3.3. Acoustic Wave Propagation in Layered Half Space

Seismograms can be synthesized quickly by simplifying an elastic wave field to a scalar acoustic wave field, because the wave field of acoustic waves in horizontally layered media can be easily calculated using recursive formulas as follows. For the ith layer, set d_i as its thickness, ρ_i as its density, $\alpha_i(\omega)$ as its attenuation coefficient, where $i = 0, 1, 2,\ldots,n$; and $i = 0$ means the ground. Denote the downgoing wave as $D_i(\omega)$ and the upgoing wave as $U_i(\omega)$ on the top of the ith layer, and the downgoing wave as $D_i'(\omega)$ and the upgoing wave as $U_i'(\omega)$ on the bottom of the ith layer. From Eqn (1.25), we can see that the basic solution of the 1D homogeneous acoustic wave equation is a function of the negative exponential, and the exponent is a phase function whose value equals the distance divided by velocity. Assuming that a harmonic wave with a single frequency ω is propagating in the homogeneous ith layer, the equation linking the downgoing wave D_i on the top and D_i' on the bottom should be

$$D_i'(\omega) = \left[e^{-\alpha d_i} - e^{-\frac{j\omega d_i}{c_i}} \right] D_i(\omega) = q_i(\omega) D_i(\omega) \tag{3.39}$$

Similarly, the equation of the upgoing wave U_i' on the bottom linking to U_i on the top is

$$U_i'(\omega) = \left[e^{-\alpha d_i} - e^{-\frac{j\omega d_i}{c_i}} \right] U_i(\omega) = q_i^{-1}(\omega) U_i(\omega) \tag{3.40}$$

By merging Eqn (3.39) with Eqn (3.40) and the equation of wave field vectors, the propagation matrix can be rewritten as

$$\begin{bmatrix} D_i' \\ U_i' \end{bmatrix} = \begin{bmatrix} q_i & 0 \\ 0 & q_i^{-1} \end{bmatrix} \begin{bmatrix} D_i \\ U_i \end{bmatrix} \tag{3.41}$$

On a horizontal interface, the acoustic wave field satisfies the soft boundary condition. Assume that R_i is the reflective coefficient of the ith layer and T_i is the transmission coefficient. Considering the first boundary condition on the bottom of the ith layer where $z = d_i$, that is the downgoing waves equal downgoing transmission waves plus upgoing reflection waves including multiples, we have

$$D_{i+1}(\omega) = T_i D_i'(\omega) - R_i U_{i+1}(\omega) \tag{3.42}$$

By considering the second condition on the bottom of the ith layer, that is the upgoing waves equal the upgoing reflection waves plus upgoing transmission waves from the lower layer, we have

$$U_i'(\omega) = R_i D_i'(\omega) + (1 + R_i) U_{i+1}(\omega) \tag{3.43}$$

Equations (3.41)–(3.43) show the propagation of acoustic waves in homogeneous isotropic layered media, expressed as recursive forms.

The reflection wave field $U_i(\omega)$ can be calculated in a recursive way by using Eqns (3.39)–(3.43). Considering internal-layer multiples, it is easy to compute the ratio of the upgoing waves versus that of the downgoing waves, which can be expressed as

$$Y_i(\omega) = \frac{U_i(\omega)}{D_i(\omega)} = q_i^2(\omega) \frac{R_i + Y_{i+1}(\omega)}{1 + R_i Y_{i+1}(\omega)} \tag{3.44}$$

Ignoring internal-layer multiples, the recursive formula for the ratio becomes

$$Y_i(\omega) = q_i^2(\omega) \left[R_i + (1 - R_i) Y_{(i+1)}(\omega) \right] \tag{3.45}$$

The above two equations reveal the propagation regulation of an acoustic wave field in horizontally layered media, and present the explicit forms corresponding to Eqns (3.36)–(3.38) in a particular situation. The computation of the synthetic seismogram of using this recursive procedure is very fast. If one assumes that detectors locate on the surface where $i = 0$, the calculation of a wave field starts from the bottom of the nth layer to the surface. For the lowest layer $i = n$, because there does not exist any upgoing waves, $U_n = Y_n = 0$, and

$$Y_{n-1}(\omega) = q_{n-1}^2(\omega) R_{n-1}$$

The ratio Y_1 on the surface can be obtained by recursive computation from the bottom layer to the top layer by using (3.44) or (3.45).

On the ground where $i = 0$, the wave field u_{z0} can be obtained by ignoring the multiples:

$$u_{z0}(\omega) = W(\omega) Y_1(\omega) \tag{3.46}$$

$W(\omega)$ is the vertical incident planar wavelet. Considering multiples on the surface boundary conditions, that is the vibration (including the upgoing waves and downgoing waves) above the ground equal zero, there only exist the incidence planar wave. Thus, we see only the upgoing reflection wave and downgoing multiples on the surface, resulting in

$$u_{z0}(\omega) = W(\omega) + (1 - R_0)U_1(\omega) \tag{3.47}$$

Substituting $Y_1 = U_1/D_1$ into Eqn (3.47) yields

$$u_{z0}(\omega) = (1 - R_0)\frac{Y_1(\omega)W(\omega)}{1 + R_0 Y_1(\omega)} + W(\omega) \tag{3.48}$$

This is what we need for synthetic reflection seismograms.

3.4. ELASTIC WAVES IN FLUID-SATURATED SOLID (I): GASSMANN'S MODEL

Elastic waves propagating in fluids or solids have been discussed in the previous two chapters, while seismic waves propagating in fluid-saturated rocks pose a problem in geophysics. As oil and gas exploration is carried out in sedimentary basins, we have to study the propagation of seismic waves in sedimentary rocks that usually belong to fluid-saturated porous solids. It is an important step that generalizes the elastic wave theory in solid media to wave theory in fluid-saturated porous solids. In the middle of the twentieth century, Gassmann (1951); Boit (1956, 1962), and others have made some outstanding contributions in this subject. Later on, the wave theory has been further developed and used for analyzing the acoustic logging responses in oil/gas reservoirs (White, 1976, 1987; Berryman, 1995, 2007). The theory of elastic waves in fluid-saturated porous solids becomes the basis for tracing and delineating oil/gas reservoirs. How to locate reservoir rocks and how to compute the fluid content in reservoir rocks are currently very hot topics in applied geophysics.

3.4.1. The Gassmann Model

To build a physical model, Gassmann (1951) simplified sedimentary rocks as being the combination of a solid frame and fluid-saturated aperture gaps. He supposed the simplest assumption that the solid frame is linear isotropic elastic media and that the interaction on mass motion between the frame and the aperture gaps can be ignored. Denote Φ as the porosity, which is usually <0.25. The indexes s or f denote the solid frame and porous fluid, respectively, and the density of a fluid-saturated porous solid equals the weighted mean of the two densities of the solid frame and the fluids

$$\rho = \Phi\rho_f + (1 - \Phi)\rho_s \tag{3.49}$$

The next question is how to describe the wave field. One may use pressure for fluid and displacement for solids to express the wave field. However, what will we use to express the wave field in the fluid-saturated solid? Gassmann chose pressure. If one assumes that the difference between the pressure at a certain time and at a previous time is ΔP, and that the difference in the volume of the mass element between the two different times is ΔV, then the mass element belongs to a fluid-saturated porous solid. After assuming that the motion interaction between the solid frame and porous fluid can be neglected, then one gets

$$\Delta P = \Delta P_t + \Delta P_f \tag{3.50}$$

$$\Delta V = \Delta V_s + \Delta V_f \tag{3.51}$$

where ΔV_s is the variation of the volume of the solid frame, ΔP_f is the variation of the fluid pressure, ΔP_t is the variation of the pressure acting on the solid frame, which is composed of two parts: the fluid pressure ΔP_f and the frame stress ΔP_s from the adjacent mass element.

The next question is how to describe the elastic parameters of a fluid-saturated porous solid. We can surely express fluid velocity as V_p because no S-wave propagates in a fluid. What elastic parameters can be used for the solid frame in a fluid-saturated porous solid? Gassmann chose the bulk modulus k. The definition of the bulk modulus k is the negative ratio of ΔP versus $\Delta V/V$, where V is the volume of a mass element and $\Delta P = -k(\Delta V/V)$. In the fluid-saturated porous solid, the variation of fluid pressure changes the volume of both the solid frame and the fluid apertures, that is

$$\Delta V_f = -\Phi V \Delta P_f / k_f \tag{3.52}$$

$$\Delta V_{s1} = -(1 - \Phi)V\Delta P_f / k_s; \tag{3.53}$$

In addition, the variation of fluid pressure changes the entire volume of the mass element

$$\Delta V_{s2} = -V\Delta P_t / k_t; \tag{3.54}$$

The above three formulas are the definitions of k_f, k_s, and k_t, respectively. Here k_t denotes the bulk modulus of a solid frame (without fluids) and k_s denotes the bulk modulus caused by the fluids in the solid frame. Although they are both the bulk moduli of the fluid-saturated porous solid, they have different meanings. Again, k_f denotes

TABLE 3.1   Parameters in Gassmann's Model for Water-Saturated Porous Solids

Parameters	Pore Fluid	Fluid Acting on a Solid	Solid Frame	Other Parameters
Foot index	f	t	s	Φ (Porosity)
Density	ρ_f		ρ_s	ρ
Pressure wave difference	ΔP_f	ΔP_t		ΔP
Volume difference	ΔV_f		ΔV_{s1} Caused by fluid pressure	ΔV_{s2} Caused by neighboring frame elements
Bulk modulus	K_f	K_t	k_s	k_φ Porous modulus

the bulk modulus of pore fluids and is treated as a fluid parameter in reservoir geophysics.

As there are more parameters of Gassmann's model to be involved, we show these parameters with different indices in Table 3.1 for a clearer presentation. In a fluid-saturated porous solid, the volume variation of a mass element includes three parts, the variant of the fluid volume caused by the fluid pressure, the variant of the volume of the solid frame caused by the fluid pressure, and the variant of the volume of the frame caused by the adjacent frame element, that is

$$\frac{\Delta V}{V} = -\left[\frac{\varphi}{k_f} + \frac{(1-\varphi)}{k_s}\right]\Delta P - \frac{1}{k_s}\,\Delta P_t \qquad (3.55)$$

Substituting Eqn (3.55) into the definition of the bulk modulus of k, that is

$$\Delta P = -k(\Delta V/V),$$

we can obtain the analytical expression of bulk modulus k for the entire volume of the mass element

$$k = \frac{\dfrac{\varphi}{k_s} - \dfrac{\varphi}{k_f} - \dfrac{1}{k_t} + \dfrac{1}{k_s}}{\left(\dfrac{\varphi}{k_t}\right)\left(\dfrac{1}{k_s} - \dfrac{1}{k_f}\right) - \left(\dfrac{1}{k_s}\right)\left(\dfrac{1}{k_t} - \dfrac{1}{k_s}\right)} \qquad (3.56a)$$

From all the assumptions mentioned above in this section, we know that the shear modulus μ_t in a solid frame equals the shear modulus μ of the fluid-saturated porous solid because there is no shear stress in fluids. Therefore, from the two elastic parameters k and μ, we can infer the wave velocity. Expression (3.56a) can be further simplified as

$$k = k_t + \frac{\left(1 - \frac{k_t}{k_s}\right)^2}{\left(\frac{\varphi}{k_f} + \frac{1-\varphi}{k_s}\frac{k_t}{k_s^2}\right)} \qquad (3.56b)$$

where the elastic parameter k of sedimentary rocks depends on μ_t and k_t of the solid frame, porosity Φ, and (k_t/k_S). The smaller the porosity is, the closer μ_t and k_t of sedimentary rocks are to those of the solid frame. As (k_t/k_S) indicates the ratio of the action on the solid frame from the adjacent solid frames versus the action caused by the fluid, when $(k_t/k_S) = 1$ and $\Phi = 0$, and the second term of the right-hand side of Eqn (3.56b) is equal to zero, the entire bulk modulus of the sedimentary rock $k \gg k_t$. When porosity increases until $k_s = k_t$, that is there exist breaks in the solid frame, the second term on the right-hand side of Eqn (3.56b) plays the main role for the elastic parameters of sedimentary rocks, and the entire bulk modulus k_f in sedimentary rocks plays the most important role.

The velocity of the seismic wave in fluid-saturated porous solid media can be calculated from the elastic parameters k and μ. In the case that the pore scale is far smaller than the wavelengths of seismic waves, the velocities of P- and S-waves can be expressed as

$$V_p = [(3k + 4\mu)/(3\rho)]^{1/2} \qquad (3.57a)$$

$$V_s = [\mu/\rho]^{1/2} \qquad (3.57b)$$

Gassmann's model is the tersest generation from the theory of elastic mechanics to the theory of fluid-saturated porous media, following the procedure of deducing the scalar wave equations.

3.4.2. The Generalized Gassmann Model

Gassmann assumed that the motion interaction between the solid frame and the porous fluids is negligible, leading to the loss of some important behaviors in waves. The permeability of sedimentary rocks is usually greater than zero, which means there is a relative movement between the solid frame and the porous fluids in sedimentary rocks. Inspired by Boit's theory, which will be discussed in the next section, a generalized Gassmann's model was proposed based on Gassmann's model by adding more parameters to the fluid-saturated porous solid. Denote κ as the permeability, and η as the viscosity coefficient of the porous fluids (Murphy et al., 1986; White et al., 1976, 1983). The flow velocity of the fluids through a solid frame under the pressure difference of the fluids in the direction of the x-axis is not equal to zero anymore, but

$$v(x,t) = -\frac{\kappa \partial p(x,t)}{\eta \partial x} \tag{3.58}$$

Equation (3.58) is an empirical equation from experiments, representing the relative motion rule between the solid frame and the porous fluids: the flow velocity of fluids is proportional to the differences of pressure and the permeability κ, and is inversely proportional to the viscosity coefficient η of the fluids. How to correlate the variation of the pressure wave field with the fluid motion? As the flow velocity of the fluids along pores is controlled by the variations of the pressure wave field, the movement of the fluids changes the volume of the mass element, we must link it to the bulk modulus k that is defined as $\Delta p = -k(\Delta V/V)$. White et al. (1976) proved that the quantitative relationships between the parameters v, p, and k can be expressed as

$$\frac{\partial p}{\partial t} = -k_e \frac{\partial v}{\partial x} \quad \text{or} \quad \frac{\partial v}{\partial x} = -\frac{1}{k_e}\frac{\partial p}{\partial t} \tag{3.59}$$

where k_e represents the "effective bulk modulus" of a mass element, which is expressed as

$$k_e = \left(\frac{\varphi}{k_f} + \frac{1-\varphi}{k_s} - \frac{k_t}{k_s^2}\right)^{-1}\frac{M_t}{M} \tag{3.60}$$

and

$$M = \lambda + 2\mu; \quad M_t = \lambda_t + 2\mu$$

where $M = V_p^2 \rho$, denoting the planar wave modulus of compression–shrinkage waves in the fluid-saturated porous solid media, M_t is the planar wave modulus of the compression–shrinkage wave for the solid frame in the fluid-saturated porous solid, that is the parameter for the porous solid without any fluids.

Computing the derivatives of Eqn (3.58) with respect to x, and then substituting it into Eqn (3.59), we can obtain the second-type P-wave in fluid-saturated porous solid, which had been deduced originally by Biot in 1956

$$\frac{\partial^2 p}{\partial x^2} = \frac{\eta}{\kappa k_e}\frac{\partial p}{\partial t} \tag{3.61}$$

The general solutions of Eqn (3.61) can be written as

$$p(x,t) = A\,\exp(-\alpha x)\,\exp(i\omega t), \quad \alpha = (i\omega \eta/\kappa k_e)^{1/2} \tag{3.62}$$

$$v(x,t) = (\kappa \alpha/\eta)\,A\,\exp(-\alpha)\,\exp(i\omega t) \tag{3.63}$$

where A is a constant determined by the initial condition, α is the complex attenuation factor of the second-type pressure wave. The real part of the complex attenuation factor is denoted by α_{p2}, which is the attenuation factor of the second-type pressure wave, and the imaginary part is proportional to the wave slowness of the second-type pressure wave, that is

$$\alpha = \alpha_{p2} + i\omega/V_{p2} \tag{3.64}$$

$$\alpha_{p2} = (\omega\eta/2\kappa k_e)^{1/2} \tag{3.65}$$

$$V_{p2} = (2\omega\kappa k_e/\eta)^{1/2} \tag{3.66}$$

The generalized Gassmann's model shows the existence of the second-type pressure wave caused by the relative motion between the solid frame and the porous fluids, getting results similar to Biot's theory. The attenuation α_{p2} of the second-type pressure wave is proportional to the square root of the viscosity coefficient, and the wave velocity is inversely proportional to the square root of the viscosity coefficient. In reflection seismology, the generalized Gassmann's model can be used in the interpretation of single-component reflection seismic records. As it assumes that the scale of aperture gaps in sedimentary rocks is much smaller than the wavelength of seismic waves, the generalized Gassmann's model is too simple to be used in the study of wave propagation in oil and gas reservoirs, and especially limited in the use of the sonic logging data analysis.

3.5. ELASTIC WAVES IN FLUID-SATURATED SOLID (II): BIOT'S THEORY

It was Biot who first found the second-type pressure wave in 1956. In his articles published in 1956, he deduced the wave equation in a fluid-saturated porous solid. He applied the theory of continuum mechanics and proposed the initial and boundary value conditions to the problem, analyzed the elastic wave equations for the porous solid, and revealed the propagation features of elastic waves in porous sandstones reservoirs. In his advanced work, he predicted the existence of the second-type pressure wave, and pointed out its characteristics. Later, his prediction was proved by laboratory experiments. Biot's theory is one of the most brilliant contributions in reflection seismology.

In Biot's theory, a solid frame in the fluid-saturated porous solid is formed by connected isometric grains, and the typical isometric grain is a small spheroid. From the viewpoint of microscales, the solid frame is similar to the linear isotropic elastic solid. The aperture gaps between the frame grains are filled with viscous fluids. Biot distinguished the displacement vector of solid mass elements from the average displacement of porous fluids, and described the relative movement and their wave field. The difference between the velocity of displacement of a solid mass element and that of porous fluids is called the relative velocity, which is proportional to the vibration friction generated by the motion. The friction is proportional to the coefficient of viscosity of the fluids and is expressed in a source term of the wave equation. The viscosity is a specific property of the fluid-saturated porous solid. Biot used the displacement potential of a mass element to express the strain, deducing the stress—strain relationship equation and the equation of motion equilibrium in the fluid-saturated porous solid. Furthermore, he deduced a pair of differential equations for describing the propagations of waves in solids and fluids respectively. Finally, he concluded that there exist two kinds of dilatation waves and one rotational wave in infinite fluid-saturated porous solid. The second-type dilatation wave disperses heavily, and there is a phase step section in the dispersion with frequency, whose center is called the critical frequency. Only when the viscosity coefficient of the fluids becomes very small, that is the corresponding permeability of rock was large, the dispersion and attenuation of the waves could be negligible.

3.5.1. Low-Frequency Elastic Waves in a Fluid-Saturated Porous Solid

In his first article (1956), Biot discussed a homogeneous multiporous solid whose aperture gaps were filled with Poiseuille flow. Poiseuille flow is similar to the flow of blood in vessels, as well as to the movement of brine and oil in gaps of rocks. Based on Poiseuille's theory, Biot deduced the elastic wave equation in a fluid-saturated porous solid that can only be used for a low-frequency wave field. When the wavelength of seismic waves is close to the scales of pores, Poiseuille's assumption is no longer correct, so corresponding theories are not good enough, especially for the interpretation of sonic logging results. In his second article (1962), Biot defined some new parameters and generalized the previous theory to high frequencies, presenting a high-frequency approximation equations for the wave field.

Since the forces acting on the solid frame and porous fluids are different, the stresses can be sorted into two sets: stress tensors acting on the surfaces of solid mass elements and stress tensors acting on the surfaces of porous fluid mass elements. Therefore, the number of elastic parameters increases from two to four, and they are λ, μ, Q, and R. Here λ and μ denote the Lamme's coefficients of a solid frame, and Q and R denote the elastic parameters of porous fluids. There is an additional term in the constitutive equation as well, which indicates the stress acting on the porous fluids:

$$S = Qe + R\Theta, \tag{3.67}$$

where e is the module of the strain component in the direction of the primary stress of a mass element, Θ is the divergence of the mean displacement vectors of the porous fluid element. Porous-fruid stress S is linearly related to e and Θ, as stated by the generalized Hooke's law. The divergence Θ of the fluid displacement vector denotes the surface force and is proportional to the vibration fiction between solid frames and the porous fluids.

As the stress tensor acting on the solids is expressed by Eqn (3.12), while the stress tensor S acting on the fluids includes two parts of Qe and $R\Theta$, in the 3D linear isotropic fluid-saturated porous solid, stress A can be expressed by the six components in the Cartesian coordinates plus S. The constitutive equations on seven scalar variables can be obtained by generalizing Hooke's law as follows:

$$\begin{aligned}
A_{xx} &= 2\mu E_{xx} + \lambda e + Q\Theta \\
A_{yy} &= 2\mu E_{yy} + \lambda e + Q\Theta \\
A_{zz} &= 2\mu E_{zz} + \lambda e + Q\Theta
\end{aligned} \tag{3.68}$$

$$\begin{aligned}
A_{xy} &= \mu E_{xy}, \quad A_{yz} = \mu E_{yz}, \quad A_{zx} = \mu E_{zx} \\
S &= Qe + R\Theta
\end{aligned} \tag{3.69}$$

where $E_{ij} = E_{ji}$ are the strain components, denoting that the strains occur on the six faces of the volume element, and e is the stress module of the mass element

$$e = E_{xx} + E_{yy} + E_{zz} \tag{3.70}$$

Θ is the divergence of the mean displacement vector of the fruid element:

$$\Theta = \frac{\partial u_{fx}}{\partial x} + \frac{\partial u_{fy}}{\partial y} + \frac{\partial u_{fz}}{\partial z} \tag{3.71}$$

In the equations, four elastic parameters, λ, μ, Q, and R, are involved, where λ, μ are accordant with the Lamme's coefficients in solids. It can be seen from Eqn (3.68) that there are two parts of the stress S acting on the fluids, the first one is Qe, which represents the pressure in the direction of primary stress, and denotes the coupling of the compression−dilation force of solid with the fluids. The second one is $R\Theta$, which represents the traction force of the fluids in adjacent mass elements and denotes the coupling of the fluid motion with the mass element.

Due to the existence of a relative movement between the solid frame and the fluids, their displacements are not synchronous. Defining the displacements and velocities separately may lead to theoretical topics from a microcosmic view, which we do not expect. We want to analyze the propagation rules of seismic waves in the two kinds of media from a macrocosmic view; therefore, we have to assume that the scale of the aperture gaps is far smaller than the wavelength of seismic waves. In this case, the relative motion of the fluids with respect to the solid frame is dependent on the direction of vibration of the mass element but independent of the amplitude. Therefore, there are three displacement components of the solid frame, say u_x, u_y, and u_z, while there are also three displacement components for fluids, say U_x, U_y, and U_z. By substituting all of them into the corresponding kinetic equations, Boit deduced a set of elastic wave equations. The displacement component in the x-direction is

$$\mu \Delta u_x + (\lambda + \mu)\frac{\partial e}{\partial x} + Q\frac{\partial \Theta}{\partial x} = \frac{\partial^2}{\partial t^2}(\rho_{11}u_x + \rho_{12}U_x) \tag{3.72}$$

$$Q\frac{\partial e}{\partial x} + R\frac{\partial \Theta}{\partial x} = \frac{\partial^2}{\partial t^2}(\rho_{12}u_x + \rho_{22}U_x) \tag{3.73}$$

Coefficients $\rho_{11}, \rho_{12}, \rho_{21}, \rho_{22}$ are mass coupling parameters under conditions of the inhomogeneous motion of porous fluids, physical unity is consistent with that of density, the foot index "1" denotes the solid frame, and "2" denotes the fluids. Equation (3.72) corresponds to the motion equation of a solid frame, while Eqn (3.73) corresponds to the motion equation of porous fluids. The two equations denote two kinds of wave coupling with each other, which are called the first-type P-wave and the second-type P-wave, respectively. Similar equations can be obtained by computing the displacement components in the y and z directions.

To find wave velocities, Biot first considered the case of ignoring viscous attenuation. Inserting divergence to both sides of vector equations such as Eqn (3.73) yields a set of equations of the compression−dilation elastic waves propagating in fluid-saturated porous solids as

$$\nabla^2(Pe + Q\Theta) = \frac{\partial^2}{\partial t^2}(\rho_{11}e + \rho_{12}\Theta) \tag{3.74}$$

$$\nabla^2(Qe + R\Theta) = \frac{\partial^2}{\partial t^2}(\rho_{12}e + \rho_{22}\Theta) \tag{3.75}$$

In Eqn (3.74), the elastic parameter is

$$P = \lambda + 2\mu \tag{3.76}$$

This shows once more that Eqn (3.74) represents the first-type P-wave and Eqn (3.75) represents the second-type P-wave, where Q and R are elastic parameters affecting the porous fluids.

It is interesting to note that the relative motion between the porous fluids and the solid frame is restrained in the case of $e = \Theta$, resulting in a compression–dilation wave velocity:

$$V_0 = \sqrt{\frac{P + R + 2Q}{\rho}} \tag{3.77}$$

The velocity is usually called the reference velocity of the compression–dilation wave in fluid-saturated porous solids.

Inserting *curl* to both sides of the vector equations like Eqn (3.73) yields an equation of the rotational elastic wave propagating in a fluid-saturated porous solid as

$$\mu\nabla^2\omega = \rho_{11}\left(1 - \frac{\rho_{12}^2}{\rho_{11}\rho_{22}}\right)\frac{\partial^2\omega}{\partial t^2} \tag{3.78}$$

This is for the first-type rotational wave with a velocity

$$V_0 = \sqrt{\frac{\mu}{\rho_{11}\left(1 - \frac{\rho_{12}^2}{\rho_{11}\rho_{22}}\right)}} \tag{3.79}$$

The velocity is usually called the reference velocity of the rotational wave in a fluid-saturated porous solid. Obviously, it equals the velocity of the S-wave in an elastic solid when the relative motion between the fluids and solids vanishes.

After the analysis mentioned above, Biot considered the case of the existence of viscous attenuation. He again deduced a set of equations of the compression–dilation waves in a fluid-saturated porous solid, which are similar to Eqns (3.74) and (3.75). Just like what we can expect, there exists an additional item of partial derivatives with respect to t as in

$$\nabla^2(Pe + Q\Theta) = \frac{\partial^2}{\partial t^2}(\rho_{11}e + \rho_{12}\Theta) + b\frac{\partial}{\partial t}(e - \Theta) \tag{3.80}$$

$$\nabla^2(Qe + R\Theta) = \frac{\partial^2}{\partial t^2}(\rho_{12}e + \rho_{22}\Theta) - b\frac{\partial}{\partial t}(e - \Theta) \tag{3.81}$$

where b indicates the attenuation coefficient and is related to the viscosity coefficient η, porosity Φ, and permeability ν, showing

$$b = \eta\Phi^2/\nu \tag{3.82}$$

In Eqn (3.81), the attenuation term is negative because the polarity of the second-type P-wave is opposite to that of the first-type.

3.5.2. All Frequency Elastic Waves in a Fluid-Saturated Porous Solid

The dispersion of seismic waves in a fluid-saturated porous solid represents the physical interaction between solids and fluids, which vary with frequencies. In a low-frequency band, the vibration of the fluids is slow, and its amplitude is large, and its actions on the solid frame are mainly the viscous resistance and consumption of the vibration energy. In a high-frequency band, the vibration of the fluids is fast and its amplitude is small, and the viscous resistance acting on the solid frame from the porous fluids is very small, and can even promote the vibration of the solid frame. Its energy consumption can also be negligible. There is a step frequency called the critical frequency between the low-frequency bandwidth and the high-frequency one. Boit deduced the critical frequency f_c as

$$f_c = \frac{\eta \varphi}{2\pi \rho_f \kappa} \tag{3.83}$$

where ρ_f is the density of porous fluids.

The results of experiments show that the critical frequency f_c is also dependent on the scales of the pores. For the first-type P-waves in solid frames and porous fluids, velocity dispersion is weak. However, dispersion of the attenuation coefficient for the first-type P-waves is strong. For the second-type P-wave, the critical frequency f_c is located in the range of 1−5 kHz for unconsolidated water-saturated sandstones. In this case, the dispersion and attenuation do not have serious effects in reflection seismology but strongly affect sonic logging data. When the critical frequency f_c is in the range of 10−1000 kHz, the dispersion and attenuation strongly affect both reflection seismology and sonic logging. The second-type compression−dilation waves in a fluid-saturated porous solid might provide very important information for studying oil/gas reservoirs.

In this section, we just introduce the results without providing the deductions of the equations in detail. Since there exists a relative motion between a solid frame and a porous fluid, and their displacements are not synchronous. One may use a new parameter for P-wave wave velocity, which is the background velocity V_c by assuming no relative motions exist between the porous fluids and the solid frame. Marking the phase velocity of the first-type P-wave as V_I, and that of the second-type as V_{II}, the group velocity of the first-type P-wave as V_{Ig}, and that of the second-type as V_{IIg}, Biot deduced that

$$\frac{V_c}{V_{Ig}} = \frac{d}{d\left(\frac{f}{f_c}\right)} \left[\frac{f}{f_c} \frac{V_c}{V_I}\right] \tag{3.84}$$

$$\frac{V_c}{V_{IIg}} = \frac{d}{d\left(\frac{f}{f_c}\right)}\left[\frac{f}{f_c}\frac{V_c}{V_{II}}\right] \tag{3.85}$$

where f is the seismic frequency, f_c is the critical frequency shown in Eqn (3.83). Biot did not only predict the existence of the second-type P-wave but he also pointed out that the reflection energy on an interface within the fluid-saturated porous solid comprises three kinds of waves: the first-type P-wave, the second-type P-wave and the first-type S-wave.

The Gassmann's and Boit's models mentioned above can be applied to describe clastic rocks, such as sandstones, but are not good to describe limestone, which contains irregular cracks and fractures. The scattering theory for irregular pores and fractures was proposed by Kuster and Toksöz (1974). In the Kuster–Toksöz model, the solid frames are an infinitely homogeneous elastic solid and the ellipsoidal pores are filled with fluids; then the propagation of waves are described according to the scattering theory. This model is closer to suspending mass models. Berryman (1995, 2007) conducted a study on a more complicated model that contains pores, fractures, and not fully saturated fluids, which is called the double porosity model. The general wave propagation features and parameters in the double porosity rocks have been be gradually revealed, building some foundations for seismic delineation of oil and gas reservoirs.

It is a universal rule in continuum mechanics that a new boundary condition leads to new waves that propagate in different ways, making seismic exploration more powerful. As the pores and cracks contain new boundaries, further studies on the propagation of new wave phases generated by pores and cracks might be expected. At present, seismologists face the following problems: whether there exist some new seismic wave phases that propagate in oil and gas reservoirs and whether they can be used in oil and gas production to meet the requirement of the fast-developing society.

3.6. TRACKING RESERVOIRS WITH THE GASSMANN MODEL

After discussing some theoretical problems on seismic waves propagating in sedimentary rocks, we may further extend this theory to some interesting problems in applications. In reservoir engineering, one of the problems on oil production currently is to improve oil and gas products. The dynamic exploitation of the reservoirs depends on information about reservoir parameters, including spatial distributions, saturation, porosity, and permeability. Although the information can be abstracted from sonic logging data that is observed in wells, the information between wells or outside of a well could be obtained only by the study of reflection records. In the 1980s, the term "reservoir geophysics" was introduced to scientists and engineers, and it became a new branch of applied geophysics.

During oil/gas production, the state parameters, reservoir parameters, and mineral components inside the reservoirs vary with space and time. Therefore, monitoring the variations of these parameters constantly and modifying the production schemes in time are necessary to improve the recovery efficiency and to maintain product stability. Furthermore, reservoir geophysics is also one of the important strengthening methods for monitoring the variations of reservoir parameters. Different from seismic exploration, reservoir geophysics serves reservoir engineering. Of course, it is developed from seismic exploration and is closely related to the techniques developed in seismic exploration. However, the aim of reservoir geophysics is reservoirs of oil and gas, but not underground structures or traps. So it needs stricter theory and more advanced techniques compared with those of seismic exploration (White, 1987; Murphy et al., 1986).

Reservoir geophysics can be used to extract some seismic parameters of strata from seismic data, such as wave velocities, impedance, and attenuation coefficients. While the study on rock physics has proved that there exists some correspondence between seismic parameters and reservoir parameters, that is fluid content, clay content, porosity, and permeability. Therefore, seismic data can present some information for describing the spatial distribution of reservoir parameters together with measurements of rock physics and well logging. In the performance of reservoir geophysics, there are different geophysical tools in use, such as multicomponent crosshole seismic tomography, multioffset VSP and high-resolution 4D reflection seismic exploration, and geophysical logging as well. Combining the information obtained from these tools and geological observations, reservoir geophysics can be used to describe the location and many physical parameters of reservoirs for petroleum reservoir engineering, and then improve the production. The theory of seismic waves in a fluid-saturated porous solid provides one of the most important bases in reservoirs geophysics.

As a possible scheme for tracking oil and gas reservoirs, we will give a special example in this section. In Section 3.4, we have discussed Gassmann's model and obtained some formulas for computing the parameters of the model with the water-saturated porous rocks. Parameters with different subscripts are shown in Table 3.1, where k_φ indicates the elastic bulk modulus of pores. Now this parameter can be used here, and its corresponding parameter k_s indicates the bulk modulus of a solid mineral frame. k_t indicates the elastic bulk modulus of rocks without the fluids inside. Therefore, we have

$$\frac{1}{k_t} = \frac{1}{k_s} + \frac{\varphi}{k_\varphi} \tag{3.86}$$

Denoting b_{pore} as the volume of a pore in a mass element, and from the definition of bulk modulus, we obtain

$$\frac{1}{k_\varphi} = \frac{1}{b_{pore}} \frac{\partial b_{pore}}{\partial p} \tag{3.87}$$

where p is the acoustic pressure wave field. If porosity φ is not large enough, for example, $<15\%$, k_φ is almost proportional to k_s. In this case, $k_\varphi = \alpha k_s$, and Eqn (3.86) can be written as

$$\frac{1}{k_t} \approx \frac{1}{k_s}\left(1 + \frac{\varphi}{\alpha}\right) \tag{3.88}$$

Substituting Eqns (3.86)−(3.88) into Gassmann's model Eqn (3.56), the entire bulk modulus k in the water-saturated porous solid media satisfies the equation

$$\frac{1}{k} = \frac{1}{k_s} + \frac{\varphi}{k_\varphi + \frac{k_f k_s}{k_s - k_f}} \tag{3.89}$$

and

$$\frac{1}{k} \approx \frac{1}{k_s} + \frac{\varphi}{k_\varphi + k_f} \tag{3.90}$$

In Eqns (3.89) and (3.90), the second terms on the right-hand side are the most important for tracking oil and gas reservoirs, which contain parameters of porosity φ and porous fluid parameter k_f. Their variation will affect the velocity because

$$V_p = \sqrt{\frac{k + \frac{4}{3}\mu}{\rho}} \tag{3.91}$$

This is the right point that the reflection records can be used for tracking oil and gas reservoirs, because reflection seismology can be used to find V_p, and Eqn (3.91) can be used to compute the bulk modulus k.

The simplest procedure to trace oil or gas reservoirs is by calculating the ratio of the bulk modulus $\gamma = k/k_s$ that indicates rock lithology. By substituting Eqn (3.89) into Gassmann's model expressed by Eqn (3.56) and eliminating k_φ, we can obtain a transitive parameter

$$\beta \equiv \frac{k}{k_s - k} = \frac{k_t}{k_s - k_t} + \frac{k_f}{\varphi(k_s - k_f)} \tag{3.92}$$

and the ratio of the bulk modulus becomes

$$\gamma = \frac{\beta}{1+\beta} \tag{3.93}$$

Parameters such as porosity φ and modulus k_s can be obtained from well loggings. We may use acoustic well logging results as the initial parameters for V_p, V_s, and density that are expressed with superscripts "(1)" hereafter. By combing Eqns (3.91)–(3.93), we can calculate the parameter $k(x,y)$ and the ratio of bulk modulus $\gamma(x,z) = k/k_s$ using the velocity functions $V_p(x,z)$ and $V_s(x,z)$ between wells.

A seven-step procedure is demonstrated for computing the fluid parameters φ and k_f as follows:

I. Compute the initial bulk modulus and shear modulus according to their definitions:

$$k^{(1)} = \rho^{(1)} \left\{ \left[V_p^{(1)} \right]^2 - \frac{4}{3} \left[V_s^{(1)} \right]^2 \right\}; \quad \mu^{(1)} = \rho^{(1)} \left[V_s^{(1)} \right]^2 \tag{3.94}$$

II. Compute $\beta^{(1)}$ and $\gamma^{(1)}$ using Eqn (3.92):

$$\beta^{(1)} = \frac{k^{(1)}}{k_s - k^{(1)}}, \quad \gamma^{(1)} = \frac{\beta^{(1)}}{1 + \beta^{(1)}} \tag{3.95}$$

III. For an arbitrary crosswell mass element j, compute density $\rho^{(j)}$ from $V_p^{(j)}$ according to the statistical relationship between the velocity of the P-wave and the density measured from rock physics.

IV. Compute the shear modulus $\mu^{(j)}$:

$$\mu^{(j)} = \rho^{(j)} \left[V_s^{(j)} \right]^2 \tag{3.96}$$

V. From the definition of the bulk modulus and Eqn (3.96), obtain the bulk modulus between wells by

$$k^{(j)} = \rho^{(j)} \left\{ \left[V_p^{(j)} \right]^2 - \frac{4}{3} \left[V_s^{(j)} \right]^2 \right\}; \tag{3.97}$$

VI. Compute $\beta^{(j)}$, $\gamma^{(j)}$ according to Eqn (3.95):

$$\beta(j) = \frac{k^{(j)}}{k_s - k^{(j)}} \quad \text{and} \quad \gamma^{(j)} = \frac{\beta^{(j)}}{1 + \beta^{(j)}} \tag{3.98}$$

VII. Because

$$\beta^{(j)} - \beta^{(1)} = \frac{k_f^{(j)}}{\varphi \left(k_s - k_f^{(j)} \right)} - \frac{k_f^{(1)}}{\varphi \left(k_s - k_f^{(1)} \right)} \tag{3.99}$$

Solve Eqn (3.99) to compute φ or the parameter of the porous fluids k_f for the jth element. Go back to step III for the next mass element between adjacent wells.

In the procedure mentioned above, we must have the velocity functions $V_p(x,z)$ and $V_s(x,z)$ first by using some seismic inversion procedures that will be discussed in Chapter 7. In addition, we have to use some parameters provided by well logging that will be taken as the benchmarks. Next, we substitute all these parameters into Gassmann's model and compute fluid parameter $k_f(x,z)$ and the ratio of the bulk modulus $\gamma(x,z)$. The theory of the direct inversion of the reservoir parameters from the seismic data has not been mature now, and more discussion will be given at the end of Chapter 7.

Wave Equation Reduction with Reflection Seismic Data Processing

Since the 1950s, reflection seismic data processing has been an indispensable branch of science included in the petroleum exploration industry. Many people think that the goal of seismic data processing is to improve the resolution and signal-to-noise (S/N) ratio of reflection seismic data. From the viewpoint of wave equations, the variation of the data flow in reflection data processing, which is the essence of seismic processing, is actually the progressive simplification of the solutions of wave equation boundary value problems. Wave equations can be finally simplified to the acoustic wave field corresponding to zero initial conditions and with boundaries located at reflectors.

Reflection Seismology
http://dx.doi.org/10.1016/B978-0-12-409538-0.00004-X

The goals of reflection data processing involve the following four aspects: (1) to standardize the seismic data format and reduce its dimension; (2) to remove the waves created by unnecessary reflectors, such as the ground surface and seabed; (3) to turn the data of elastic wave equations to data that satisfy the pure acoustic wave equation; and (4) to obtain the 3D wave data volume in which reflection waves match the underground reflector and reduce any distortions at the same time. In other words, the final data set produced by data processing should present the images of "soft boundaries" associated with the wave problem under study.

In this chapter, the variation of the wave equations synchronously with the data flow in seismic processing will be analyzed mainly in two-dimensional (2D) cases, to demonstrate the progressive simplification of the boundary value problems of seismic wave equations.

4.1. THE STATICS OF LAND SEISMIC DATA

We consider 2D land seismic data acquisition on point sources and single-component receivers that are located in unconsolidated surface soils of low velocity. The reflection energy of the upgoing wave is mainly represented by its vertical component, although there is some energy of shear waves in the soil layer. The signal recorded with single-component receivers can be represented by the vertical component of the displacement wave $w(x,y,z,t)$. Because of the heterogeneity of the surface layers, there are noises included in the records caused by surface relief and by the poor coupling between the soil and geophones, resulting in a lower S/N ratio compared with that in offshore seismic data. The single-component seismic records on land can be described as the elastic wave Eqn (4.1), that is

$$(\lambda + 2\mu)\frac{\partial^2 w}{\partial z^2} + \mu\left(\frac{\partial^2 w}{\partial x^2} + \frac{\partial^2 w}{\partial y^2}\right) + (\lambda + \mu)\left(\frac{\partial^2 v}{\partial y\partial z} + \frac{\partial^2 u}{\partial x\partial z}\right) - \rho\frac{\partial^2 w}{\partial t^2}$$
$$= W(t)\delta(x_s, y_s, z_s) \tag{4.1}$$

where $W(t)$ is the wavelet function of the point source located at co-ordinates (x_s, y_s, z_s). If one assumes that the partial derivative on y is zero, the 2D single-component seismic data can be approximated as

$$(\lambda + 2\mu)\frac{\partial^2 w}{\partial z^2} + \mu\frac{\partial^2 w}{\partial x^2} + (\lambda + \mu)\frac{\partial^2 u}{\partial x\partial z} - \rho\frac{\partial^2 w}{\partial t^2} = W(t)\delta(x_s, z_s) \tag{4.2}$$

Owing to the transverse components u and v not being observed and as there is coupling between u, v, and the orthogonal component w, we must perform special processing to eliminate the coupling effect of components

u, v by using only single-component land seismic data w (see Section 4.3 for details).

The wave field of land 3C seismic data consists of three displacement components u, v, and w, where w usually has a higher S/N ratio and v has a lower S/N ratio. The 3C land seismic data can be approximately described by the following elastic wave equation:

$$\frac{\partial^2 u}{\partial t^2} = \frac{\partial}{\partial x}\left[c_p^2\left(\frac{\partial u}{\partial x} + \frac{\partial v}{\partial y} + \frac{\partial w}{\partial z}\right)\right] - 2\frac{\partial}{\partial x}\left[c_s^2\left(\frac{\partial v}{\partial y} + \frac{\partial w}{\partial z}\right)\right]$$

$$+ \frac{\partial}{\partial y}\left[c_s^2\left(\frac{\partial u}{\partial y} + \frac{\partial v}{\partial x}\right)\right] + \frac{\partial}{\partial z}\left[c_s^2\left(\frac{\partial w}{\partial x} + \frac{\partial u}{\partial z}\right)\right]$$

$$\frac{\partial^2 v}{\partial t^2} = \frac{\partial}{\partial y}\left[c_p^2\left(\frac{\partial u}{\partial x} + \frac{\partial v}{\partial y} + \frac{\partial w}{\partial z}\right)\right] - 2\frac{\partial}{\partial y}\left[c_s^2\left(\frac{\partial u}{\partial x} + \frac{\partial w}{\partial z}\right)\right]$$

$$+ \frac{\partial}{\partial z}\left[c_s^2\left(\frac{\partial v}{\partial z} + \frac{\partial w}{\partial y}\right)\right] + \frac{\partial}{\partial x}\left[c_s^2\left(\frac{\partial u}{\partial y} + \frac{\partial v}{\partial x}\right)\right]$$

$$\frac{\partial^2 w}{\partial t^2} = \frac{\partial}{\partial z}\left[c_p^2\left(\frac{\partial u}{\partial x} + \frac{\partial v}{\partial y} + \frac{\partial w}{\partial z}\right)\right] - 2\frac{\partial}{\partial z}\left[c_s^2\left(\frac{\partial u}{\partial x} + \frac{\partial v}{\partial y}\right)\right]$$

$$+ \frac{\partial}{\partial x}\left[c_s^2\left(\frac{\partial w}{\partial x} + \frac{\partial u}{\partial z}\right)\right] + \frac{\partial}{\partial y}\left[c_s^2\left(\frac{\partial v}{\partial z} + \frac{\partial w}{\partial y}\right)\right]$$

(4.3)

The first boundary condition involved with Eqns (4.1)–(4.3) is the surface between the air and the solid earth. The transverse displacement components u and v, as well as stress, are discontinuous while the vertical displacement component w is continuous:

$$[A_{zx}] = 0, \quad [A_{zy}] = 0 \quad [A_{zz}] = 0; \quad [w] \approx 0_o \qquad (4.4)$$

where [*] denotes the difference between the boundary values by one side and another, A denotes the second derivatives of the displacement components with respect to the subscripts. The boundaries defined by Eqn (4.4) are called "hard boundaries", which produce strong reflections. Between solid rock layers underground, all the displacement and stress components are continuous, that is

$$[A_{zx}] = 0, \quad [A_{zy}] = 0 \quad [A_{zz}] = 0; \quad [u] = 0, \quad [v] = 0, \quad [w] \approx 0_o \qquad (4.5)$$

The boundaries defined by Eqn (4.5) are called "soft boundaries", which usually correspond to underground reflectors.

The statics process is a tool for the dimensional reduction of seismic data to eliminate the impact by elevation coordinates (Sheriff, 1983, 1984; Telford et al., 1990). In the acquisition of 3D/3C land seismic records, given the coordinate of source (x_s, y_s, z_s) and that of the receiver (x_r, y_r, z_r), three displacement components before statics can be expressed as a seven-dimensional function,

$$u(x_s, y_s, z_s, x_r, y_r, z_r, t)$$
$$v(x_s, y_s, z_s, x_r, y_r, z_r, t)$$
$$w(x_s, y_s, z_s, x_r, y_r, z_r, t)$$

The three displacement components after statics can be expressed as a five-dimensional function,

$$u(x_s, y_s, 0, x_r, y_r, 0, t) = u(x_s, y_s, x_r, y_r, t)$$
$$v(x_s, y_s, 0, x_r, y_r, 0, t) = v(x_s, y_s, x_r, y_r, t)$$
$$w(x_s, y_s, 0, x_r, y_r, 0, t) = w(x_s, y_s, x_r, y_r, t)$$

Thus, two dimensions of the wave field function have been reduced in the statics process. If the homogeneous elastic wave equation is approximately used for the wave field, the wave equations after statics can be written for 3C land seismic data as

$$(\lambda + 2\mu)\frac{\partial^2 u}{\partial z^2} + \mu\left(\frac{\partial^2 u}{\partial x^2} + \frac{\partial^2 u}{\partial y^2}\right) + (\lambda + \mu)\left(\frac{\partial^2 v}{\partial y \partial x} + \frac{\partial^2 w}{\partial x \partial z}\right) - \rho\frac{\partial^2 u}{\partial t^2}$$

$$= W(t)\delta(x_s, y_s, 0)$$

$$(\lambda + 2\mu)\frac{\partial^2 v}{\partial z^2} + \mu\left(\frac{\partial^2 v}{\partial x^2} + \frac{\partial^2 v}{\partial y^2}\right) + (\lambda + \mu)\left(\frac{\partial^2 w}{\partial y \partial z} + \frac{\partial^2 u}{\partial y \partial x}\right) - \rho\frac{\partial^2 v}{\partial t^2} \qquad (4.6)$$

$$= W(t)\delta(x_s, y_s, 0)$$

$$(\lambda + 2\mu)\frac{\partial^2 w}{\partial z^2} + \mu\left(\frac{\partial^2 w}{\partial x^2} + \frac{\partial^2 w}{\partial y^2}\right) + (\lambda + \mu)\left(\frac{\partial^2 v}{\partial y \partial z} + \frac{\partial^2 u}{\partial x \partial z}\right) - \rho\frac{\partial^2 w}{\partial t^2}$$

$$= W(t)\delta(x_s, y_s, 0)$$

In cases of the single-component acquisition of land 2D seismic data, given the coordinate of source (x_s, z_s) and that of the receiver (x_r, z_r), data records can be described as a five-dimension function $w(x_s, z_s, x_r, z_r, t)$. After statics, the wave field function becomes

$$w(x_s, 0, x_r, 0, t) = w(x_s, x_r, t)$$

The wave equation after statics is

$$(\lambda + 2\mu)\frac{\partial^2 w}{\partial z^2} + \mu\frac{\partial^2 w}{\partial x^2} + (\lambda + \mu)\frac{\partial^2 u}{\partial x \partial z} - \rho\frac{\partial^2 w}{\partial t^2} = W(t)\delta(x_s, 0) \qquad (4.7)$$

The statics discussed above does not completely eliminate the effect of landscape reliefs on seismic waves. As a process for dimension reduction, statics only aims at eliminating the impact of elevation coordinates on the wave phase and cannot completely remove the wave distortion and echoes caused by the land reliefs. Continuation methods that extrapolate the wave field from a land relief into a plane can do better for removing all the effects coming from landscape reliefs (see Chapter 6 for details).

4.2. MUTING AND DEGHOST FILTERING

Muting is a special process of filling zeros into the seismic data corresponding to the direct and head waves in a common-source gather. If one takes offshore seismic data for instance, the pressure field $u(x,0,t)$ satisfies the acoustic wave equation and $u(x,0,t) = u_1(x,0,t) + u_2(x,0,t)$, where $u_2(x,0,t)$ satisfies the acoustic equation with varied coefficients and involves all the reflection waves. The direct and head waves $u_1(x,0,t)$ satisfy the acoustic equation with constant coefficients as follows:

$$\frac{\partial^2 u_1}{\partial x^2} + \frac{\partial^2 u_1}{\partial z^2} - \frac{1}{c_1^2}\frac{\partial^2 u_1}{\partial t^2} = -W(t)\delta(x_s, z_s) \tag{4.8}$$

where (x_s, z_s) denotes the coordinates of the point source, and c_1 is the P-wave velocity of sea water. The muting process (Sheriff, 1983, 1984; Telford et al., 1990) uses wave records $u(x,0,t)$ as the input to produce $u_2(x,0,t)$ as the output. It can be implemented through an interactive display of the prestack seismic data and by cutting the direct waves and head waves by operators, and then filling zeros from the start moment to those phases on the seismogram.

The significant signals received in reflection records are mainly upgoing reflection waves generated by underground reflectors. So, the direct waves and head waves with strong energy should be muted, to visually enhance the reflection waves with weak energy. Although the direct and head waves are to be muted, the reflection generated by the low-velocity surface layer cannot be completely removed by the muting process, and we still need deghost filtering to remove the ghost wave.

The ghost wave emerges when sources are not located deep enough, so the surface reflective is attached to the excitation signals, making the excitation signal undesirable. A desirable signal on reflection seismograms should be pulse-like. For example, when seismic waves are produced in shot depth beneath the surface or seawater, the source signals and their strong reflective waves from the surface (namely the ghost wave) are added up and attach to the direct wave and downward to underground. Thus, all the underground reflection waves are affected by the ghost wave that cannot be eliminated in the muting process.

If we denote the wavelet function as $W(t)$ in the source items of wave Eqns (4.7) and (4.8), the reflection wave field without surface boundaries is $u(x,0,t)$, and that with surface boundaries is $u'(x,0,t)$; the deghost process uses $u'(x,0,t)$ and $W(t)$ to obtain $u(x,0,t)$. Let the wavelet function with the ghost wave be $W'(t)$. Then, both $W(t)$ and $W'(t)$ can be recorded after surface tests; a linear filter can be employed for the deghost process, which belongs to a kind of deconvolution filters and can be easily performed in the frequency domain, producing $u(x,0,t)$ as the output.

Many kinds of deconvolution filters are widely used in seismic data processing for pulse-like signals. Besides the deghost filter, the spike deconvolution filters and minimum phase deconvolution filters are also commonly used, the former aims at making the wavelet function $W(t)$ closer to a spike pulse, while the latter tries to keep the energy of the wavelet function moving ahead.

4.3. SHEAR WAVE DECOUPLING PROCESS

The shear wave decoupling process aims to remove the transversal components in single-component land seismic records. The transversal components are shown in Eqn (4.2), resulting in complex seismic data that do not satisfy the simpler acoustic wave equation. This process uses the single-component land seismic data as the input, and tries to eliminate the shear waves in the records to make the acoustic wave equations usable.

In 2D cases, the single displacement component data can be denoted as $w(x_r,t)$ for shot x_s. There have been many decoupling methods developed, including the velocity filter in the F-K domain and Tau-P filters. Based on the law that the P-wave velocity is greater than that of the S-wave, F-K filters remove all the waves with low velocity and only keep the longitudinal reflection wave in the single-component seismic records.

The F-K filter (Sheriff, 1983, 1984; Telford et al., 1990) performs the 2D Fourier transformation to a single common shot record $w(x,t)$, namely

$$w(k,\omega) = \frac{1}{2\pi} \int\limits_{-\infty}^{\infty} \int w(x,t)e^{-i(kx+\omega t)}\mathrm{d}x\,\mathrm{d}t$$

Then, multiplying it by a weight function in F-K space yields

$$H(k,\omega) = \begin{vmatrix} 1 & \text{for } |k| < \omega/V \\ 0 & \text{for } |k| \geq \omega/V \end{vmatrix}$$

where V is the threshold of the apparent wave velocity between the P-wave and S-wave velocities, which must be defined by data processing operators. Finally, the velocity filter is realized by inverse 2D Fourier transform back into the X-T domain as

$$w(x,t) = \frac{1}{2\pi} \int\int_{-\infty}^{\infty} H(k,\omega)w(k,\omega)e^{i(kx+\omega t)}\mathrm{d}k\,\mathrm{d}\omega$$

For the land seismic records, the F-K filter can usually eliminate the pure S-wave, but it is not good to eliminate other converted waves with high velocities contained in the displacement component, such as the multiple PSP-wave. So, the filtered land seismic data may still not satisfy the acoustic equation exactly. Denoting the wave equation after F-K filtering as $w(x,t)$ we can write that as

$$(\lambda + 2\mu)\frac{\partial^2 w}{\partial z^2} + \mu\frac{\partial^2 w}{\partial x^2} - \rho\frac{\partial^2 w}{\partial t^2} = W(t)\delta(x_s,0) \tag{4.9}$$

For 3D/3C land seismic data $u(x_r,y_r,0)$, $v(x_r,y_r,0)$ for S-wave and $w(x_r,y_r,0)$ for P-wave, the wave equations after F-K filtering should be

$$(\lambda + 2\mu)\frac{\partial^2 u}{\partial z^2} + \mu\left(\frac{\partial^2 u}{\partial x^2} + \frac{\partial^2 u}{\partial y^2}\right) - \rho\frac{\partial^2 u}{\partial t^2} = W(t)\delta(x_s,y_s,0)$$

$$(\lambda + 2\mu)\frac{\partial^2 v}{\partial z^2} + \mu\left(\frac{\partial^2 v}{\partial x^2} + \frac{\partial^2 v}{\partial y^2}\right) - \rho\frac{\partial^2 v}{\partial t^2} = W(t)\delta(x_s,y_s,0) \tag{4.10}$$

$$(\lambda + 2\mu)\frac{\partial^2 w}{\partial z^2} + \mu\left(\frac{\partial^2 w}{\partial x^2} + \frac{\partial^2 w}{\partial y^2}\right) - \rho\frac{\partial^2 w}{\partial t^2} = W(t)\delta(x_s,y_s,0)$$

Equations (4.9) and (4.10) might still involve some converted multiples, and accordingly have some difference with respect to the pure P-wave equation. Thus, we need more processing schedules for further eliminating the remained converted waves (see Sections 4.5 and 4.6 for details).

4.4. SUPPRESSION OF MULTIPLES GENERATED BY THE OCEAN BOTTOM

In offshore seismic data acquisition, point sources and receivers are all located in sea water, which acts like a filter to eliminate shear waves. So, the offshore seismic data record only scalar pressure field $u(x,y,z,t)$ exited by the point sources. Although there are secondary shear waves in the rock layers beneath, it becomes very weak after the filtering of sea water and can be treated as noise. The common shot offshore seismic records after degoast filtering can be defined by the acoustic Eqn (1.19), that is

$$\left[\frac{\partial^2 u}{\partial x^2} + \frac{\partial^2 u}{\partial y^2} + \frac{\partial^2 u}{\partial z^2}\right] - \frac{1}{c^2(x,y,z)}\frac{\partial^2 u}{\partial t^2} = -W(t)\delta(x_s, y_s) \qquad (4.11)$$

where $W(t)$ is the wavelet function of the point source located at (x_s, y_s). The initial condition for the equation is $u(x,y,z,t) = 0$ when $t < 0$.

For the boundary conditions, we use square brackets to denote the difference in stress or displacement components between the two sides of a boundary. In the case of offshore sedimentary basins, the seabed is the closest reflection boundary to the sources and receivers. Multiples of strong reflection from the seabed yield severe interference to primary reflections coming from deep layers, and should be suppressed in seismic data processing. For sea water/solid rock boundaries, only the longitudinal stress and displacement components are continuous, while the transversal displacement component is discontinuous. Thus, the boundary conditions should be

$$[A_{zz}]_{i=1} = 0, \quad [w]_{i=1} \approx 0, \qquad (4.12)$$

where $i = 1$ means that the seabed is the first boundary in producing multiples. The solid rock boundaries under the seabed are the main targets of seismic prospect, where both the longitudinal and transversal stress and displacement components are continuous.

$$[A_{zx}]_{i>1} = 0, \quad [A_{zy}]_{i>1} = 0, \quad [A_{zz}]_{i>1} = 0; \quad [u]_{i>1} \approx 0, \quad [v]_{i>1} \approx 0, \quad [w]_{i>1} \approx 0, \qquad (4.13)$$

where $i > 1$ means that the boundary is under the seabed.

The goal of the multiple suppression is to suppress the multiples caused by the seabed in a common-source gather record. Assume that the wave field $u(x,0,t)$ of a single shot satisfies the acoustic Eqn (4.11) and boundary conditions Eqns (4.12) and (4.13). Let $u_2(x,0,t) = u(x,0,t) - u_1(x,0,t)$ be the output of the multiple suppression process, where $u_1(x,0,t)$ must satisfy

$$\frac{\partial^2 u_1}{\partial x^2} + \frac{\partial^2 u_1}{\partial z^2} - \frac{1}{c(x,y,z)^2}\frac{\partial^2 u_1}{\partial t^2} = -W(t)\delta(x_0, z_0) \qquad (4.14)$$

and the boundary condition Eqn (4.12) without the existence of Eqn (4.13). Note that the input $u(x,0,t)$ and the output $u_2(x,0,t)$ in the multiple suppression process satisfy the same equation but with different boundary conditions; here $u_2(x,0,t)$ has nothing to do with the seabed boundaries Eqn (4.12). The suppression can be performed as follows: (1) interactively display prestack gathers and distinguish the seismogram of the seabed reflection; (2) then extract wavelet functions and compute the reflection coefficient of the seabed; and (3) use predictive deconvolution method to suppress the multiples (Sheriff and Geldart, 1983).

As the seabed is the first reflection interface in offshore sedimentary basins, after suppressing multiples the output seismic data $u_2(x,0,t)$ still contain interbed multiples coursed by solid rock layers beneath the seabed. These interbed multiples satisfy Eqn (4.11), but together with the boundary condition Eqn (4.13). As discussed in Section 3.3.3, the reflection wave field can be computed synthetically for a horizontally layered medium under two assumptions. The first is the wave field containing interbed multiples, and the second is where the wave field is without multiples. If the thickness and wave velocities of the layers are known, seismic data with and without interbed multiples can be synthesized by the formula shown in Section 3.3.3. Thus, one can compute the interbed multiples for the suppression processing. If the thickness and wave velocities of each layer are unknown, then the stack of the common-midpoint (CMP) gather is also helpful for multiple suppression.

4.5. CMP STACKING

Reflective seismic data processing has become an industry developed together with computer science. The main technical breakthrough came in the 1960s, due to the application of techniques called normal move out (NMO) and the stack of CMP gathers (Sheriff, 1983, 1984; Telford et al., 1990). By applying them, one can achieve the following goals:

1. Obtain new wave field data sets with the same dimension as that of the field acquisition, namely zero-offset gathers that indicate underground soft boundaries.
2. Enhance the energy of the primary reflections coming from the soft boundaries.
3. Attenuation of nonprimary wave energy to some degrees, making waves in the zero-offset gathers satisfy the acoustic equation and the corresponding boundary conditions.

Since the 1960s, NMO and stacking of CMP gathers have become conventional techniques in seismic data processing, and the output, called the stacked seismic section, became one of the key results of seismic processing. In the twenty-first century, although the NMO and CMP stacking seem likely to be substituted by the prestack migration procedures, they may be still useful in some cases. Before the discussion of NMO and stacking, we need to define the three seismic gathers: common shot point (CSP) gathers, common receiver point (CRP) gathers, and the CMP gathers.

Given shot point coordinates as $(x_s, y_s, z_s = 0)$, the CSP gather after statics can be denoted as $u(x_r, y_r, 0, t)$, where (x_r, y_r) are the coordinates of point receivers in the gather. Original field record data contain abandoned CSP gather that have many channels. The CSP gather discussed here belongs to a single shot gather without abandoned channels and being processed by the statics.

Given the receiver coordinates $(x_r, y_r, z_r = 0)$, the CRP gather is denoted as $u(x_s, y_s, 0, t)$, which is the mirror of the CSP gather when the image sources are placed at the location $(x_r, y_r, z_r = 0)$ and the image receivers at the location $(x_s, y_s, z_s = 0)$.

Given the coordinates of the midpoint $(x_m, y_m, z_m = 0)$, where $x_m = (x_r + x_s)/2$, $y_m = (y_r + y_s)/2$. The CMP gather is denoted as $u(x_m, y_m, 0, t)$. The shots and receivers are symmetrical in CMP gather after statics, making seismic data processing much easier.

Following the definitions stated above, the acoustic wave equations for these three gathers are different. In 2D cases, the wave field of the CSP gathers satisfies the acoustic equation

$$\left[\frac{\partial^2 u(x, z, t)}{\partial x^2} + \frac{\partial^2 u(x, z, t)}{\partial z^2}\right] - \frac{1}{c^2(x, z)}\frac{\partial^2 u(x, z, t)}{\partial t^2} = -W(t)\delta(x - x_s, z - z_s)$$

(4.15)

For the CRP gathers, we exchange the coordinates of shots and receivers in Eqn (4.15), then the acoustic equation of the CRP gather becomes

$$\left[\frac{\partial^2 u(x, z, t)}{\partial x^2} + \frac{\partial^2 u(x, z, t)}{\partial z^2}\right] - \frac{1}{c^2(x, z)}\frac{\partial^2 u(x, z, t)}{\partial t^2} = -W(t)\delta(x - x_r, z - z_r)$$

(4.16)

Let the coordinates of the midpoint be located at the origin, that is at $(x_m = 0, z_m = 0)$, the acoustic equation of the CMP gather is

$$\left[\frac{\partial^2 u(x_m, z, t)}{\partial x^2} + \frac{\partial^2 u(x_m, z, t)}{\partial z^2}\right] - \frac{1}{c^2(x, z)}\frac{\partial^2 u(x_m, z, t)}{\partial t^2}$$
$$= -W(t)\delta(x - x_{-m}, z - z_m)$$ (4.17)

where $(x_{-m}, 0)$ is a set of shot points symmetrical with receivers $(x_m, 0)$ to the Z-axis, and (x_m, z) denotes any point in the lower half space in seismic profiles.

The CMP stacking is a transformation process that transforms data satisfying 2D wave Eqn (4.17) into data satisfying a 1D acoustic wave equation. It uses longitudinal displacement component $u(x_m, 0, t)$ as the input, which satisfies 2D wave Eqn (4.17) approximately.

The expected output $u'(x', 0, t)$ satisfies the typical 1D acoustic wave equation

$$\frac{\partial^2 u'(x', t)}{\partial x^2} - \frac{1}{c^2(0, z)} \frac{\partial^2 u'(x', t)}{\partial t^2} = -W(t)\delta(x - x') \tag{4.18a}$$

where $u'(x', 0, t)$ is the longitudinal displacement component, x' is the coordinate of the midpoint, and shots and receivers are located at the same spot $(x', 0)$, forming the output seismic sections like zero offset, that is combined exciting and receiving. The stacked seismic sections are also called zero-offset sections. It will be proved in Section 5.2 that the primary reflection wave in a CMP stacking trace can be represented by a form of convolution integral.

Performance of the CMP stacking includes the following three steps: the velocity analysis, NMO together with residual statics, and horizontal stacking. Theoretically, the CMP stacking can give desirable results only for horizontally layered homogeneous media. The dimension reduction of the wave equation is restricted to the cases where the velocity function of the media varies very slowly and transversely, that is $c(x', z) = c(0, z)$. As a matter of fact, the CMP stacking works well for gently dipping formations with a dip $<15°$, but it may cause miner distortions.

The first work in the CMP stacking aims at the computation of wave velocity $c(0,t)$, where t denotes the two-way travel time and $c(0,t)$ denotes the interval velocity between interfaces of each layer under the midpoint. After velocity analysis, we can calculate the time difference between a primary reflection appearing in each channel and that appearing in the midpoint channel. The NMO processing removes out the time difference effect, aligning the primary waves horizontally in a CMP gather, making waves in each channel satisfy the 1D acoustic wave equation. Then, the horizontal stacking is performed, to add a wave field in all channels together. Let $x = x' = x_m$. Then, the resulting seismic data satisfy the 1D homogeneous wave equation as

$$\frac{\partial^2 u'(x', t)}{\partial x^2} - \frac{1}{c^2(0, z)} \frac{\partial^2 u'(x', t)}{\partial t^2} = 0 \tag{4.18b}$$

This equation equally represents a plane wave u' incident from infinity $z = -\infty$. Here, we assume that the medium above the surface would be filled by a layer similar to the first underground layer, so that the ground boundary is omitted.

We should pay a little attention to the source of the wave equation here, though it has been discussed in Section 3.1. Seismic sources are artificial vibrations located on the surface and can be mathematically defined by a source function. The seismic sources include point sources, dipole seismic source, etc., and contain band-limit frequency components. A point source function in 2D space can be expressed as

$$F(x_s, z_s, t) = \delta(x - x_s, z - z_s)W(t)$$

where (x_s, z_s) represents the spatial coordinates of point source, and $W(t)$ is the wavelet function of the limited band. The wavelet with the infinite band is described by $W(t) = \delta(t - t_0)$, which shows the unit impulse shot at time t_0. Given $t_0 = 0$,

$$F(x_s, z_s, t) = \delta(x - x_s, z - z_s)$$

defines a seismic force source to inspire a unit impulse at point (x_s, z_s). Actually, the formula above defines the initial condition of the wave field as

$$F(x, z, t) = \begin{cases} 0, & t < 0 \\ 1, & t = 0; \; x = x_s, \; z = z_s \end{cases}$$

If a point source is located at infinity, (∞, ∞), while the region under the wave field study, say Ω, is far less than infinity, then the seismic source can be seen mathematically as the plane wave incident from the outer space onto a limited underground region, given by

$$F(x_s, z_s, t) = \delta(\infty, \infty)w(t) \approx 0, \quad (x, z) \in \Omega$$

Thus, the zero source term in the wave equation is equivalent to the incident source located at infinity. The acoustic wave equation with a zero source yields the simplest wave field that becomes an ideal output in seismic data processing.

As mentioned in the last section, CMP stacking can help to suppress multiples when the exact thickness and wave velocities of each layer are not given. Actually, the stacking process enhances the reflection energy and reduces waves that cannot be realigned horizontally with the primary reflections in a CMP gather, improving the S/N ratio of seismic records. It is difficult to quantitatively evaluate the effect of the CMP stacking for multiple suppression. As discussed in Section 3.3.3, the seismic traces can be calculated in horizontally layered media with some recursive formula for both cases, that is with or without multiples. The effect of the stacking process falls between the two. In other words, multiples are partly eliminated.

Applying the CMP stacking to all CMP gathers along a profile, we obtain a CMP seismic section, which contains the combined exciting and receiving seismic traces that align reflection waves along the midpoint coordinates $(x_m, z = 0)$. Then, the 2D wave field $u(x_m, z = 0, t > 0)$ can be simplified as $u(x, t)$. What kind of equations does $u(x, t)$ satisfy? Equation (4.18b) can be rewritten as

$$\frac{\partial^2 u(x,t)}{\partial x^2} + \frac{\partial^2 u(x,t)}{\partial z^2} - \frac{1}{c^2(x,z)}\frac{\partial^2 u(x,z)}{\partial t^2} = 0, \tag{4.19}$$

with the initial condition

$$u(x,t) = 0, \quad t \leq 0$$

and the boundary condition

$$u_{\Gamma_i}^+(x',t) = u_{\Gamma_i}^-, x' \in \{\Gamma_i\}, \quad i = 1, 2, ..., I \tag{4.20}$$

where the superscript "+" stands for the upper part of the ith boundary Γ_i, and superscript "−" stands for the lower part, and I denotes the total number of boundaries. $\{\Gamma_i\}$ include all soft boundaries (Section 3.2).

4.6. THE ONE-WAY WAVE EQUATION AND THE WAVE MIGRATION EQUATIONS

Theoretically, the CMP stacking yields good results only if the wave propagates in a horizontally layered homogeneous media. In other words, dimension reduction only works when the velocity function in the coefficient of the acoustic equation satisfies the condition $c(x,z) = c(0,z)$. For steeply dipping formations and heavily relieved boundaries, there are exceptional differences between the stacked sections and real strata sections (see Figure 4.1). Poststack migration can be used to get better strata images from the stacked sections, and reduce some distortions caused by the CMP stacking (Claerbout, 1985; Berkout, 1985).

The reflective waves $u(x,t)$ after the CMP stacking are two-way (upward and downward) waves that satisfy the initial and the boundary value problem of Eqn (4.19). The 2D wave migration is a process to calculate a special function $u(x,z > 0,t)$ by the given $u(x,0,t)$ with the given wave velocity function $c(x,z)$. This process involves the backscattering propagation of the one-way wave from the recorded surface to its sources, that is the underground reflectors. That is, the one-way acoustic wave equation should be discussed first.

Because the travel time of the one-way wave (upgoing wave or downgoing wave) equals half that of the two-way wave and assuming $T = t/2$, $dT = dt/2$, one can see that Eqn (4.19) becomes

$$\frac{\partial^2 u}{\partial x^2} + \frac{\partial^2 u}{\partial z^2} - \frac{4}{c^2(x,z)}\frac{\partial^2 u}{\partial T^2} = 0 \tag{4.21}$$

This is the one-way wave equation, and is also called a wave migration equation in applications, where T is the one-way traveling time. Assume

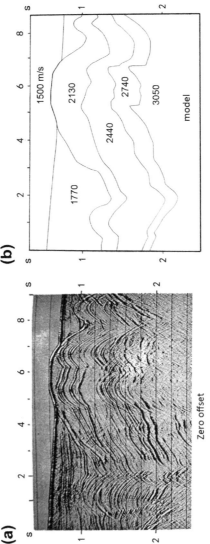

FIGURE 4.1 Comparison between the stacked $u(x,t)$ section (a) and real strata $v(x,t)$ model (b). *Source: James (1989).*

V to be the half wave velocity function, $V = c(x,z)/2$, the equation above can be rewritten as

$$\frac{\partial^2 u}{\partial x^2} + \frac{\partial^2 u}{\partial z^2} - \frac{1}{V^2(x,z)}\frac{\partial^2 u}{\partial t^2} = 0, \tag{4.22}$$

which means that the one-way wave equation is equivalent to the two-way wave equation if the media velocity is reduced by half.

After the CMP stacking, the zero-offset 2D wave field $u(x, z = 0, t > 0)$ belongs to the wave field observed at the surface. We want to get the images of the reflectors that are buried underground far away from the receivers. Wave propagation from reflectors to the receivers causes diffusion and distortion in the images shown on the stacked sections. If the ground wave field propagates back to the reflectors, it can give good images of the reflectors, eliminating the distortions. This is the goal of the wave migration process. The migration process uses the stacked section as the input to compute the reflecting wave field $u(x, z, \tau)$ that is newly created very close to the reflectors, showing better images of the reflectors. Besides, two-way time t is not a good indicator of the depth that is also dependent on velocity. Considering the one-way backpropagation of the wave field to underground reflectors, we rewrite the approaching time lag τ by applying the half velocity as follows:

$$\tau = 2z/c(x,z) \tag{4.23}$$

The approaching time lag τ is the time difference between the wave field front and that which approaches the underground target located at any point (x,z). When waves are migrated by backpropagation to deeper reflectors with increasing τ, we obtain the correct images of the reflectors located at different depths.

Classical physicists believe in the symmetry of the universe. Since the approaching time lag τ indicates depth, there must be a corresponding approaching distance δ that indicates waves traveling along the positive time direction, and can be defined based on Eqn (4.23) as

$$\delta = c(x,z)t/2 \tag{4.24}$$

The approaching distance δ indicates the distance from the wave field front surface to the underground imaging target (x,z). When waves are migrated by backpropagation to deeper reflectors with decreasing δ, we obtain the correct images of the reflectors located at different depths. Thus, the wave migration can be considered as a process to transform the wave field received from the surface into wave signals in the $\tau - \delta$ domain by continuously making δ approach zero.

Professor Clearbout first proposed the idea of seismic wave migration and put forward the so-called Clearbout migration equation. He defined the coordinate transformation in the δ domain as

$$D = \delta + z = c(x,z)t/2 + z \quad \text{and} \quad d = z \tag{4.25}$$

where z denotes the point depth in the CMP-stacked section.

Under the transformation, the stacked section $u(x,0,t)$ can be written as

$$u(x,0,D) = u(x,0,\delta),$$

where D is the apparent depth of the reflecting point to be imaged. Because the depth d in the stacked section is different from D, there appear distortions in the stacked section, which need to be migrated. The key part for distortion elimination is the algorithm of making δ approach zero. However, the transformation stated above prerequires knowing the accurate velocity $c(x,z)$.

In the coordinates (D,d), inserting Eqn (4.25) into the one-way wave Eqn (4.21) results in the acoustic equation for $u(x,D,d)$, namely

$$\frac{\partial^2 u}{\partial x^2} + \frac{\partial^2 u}{\partial d^2} - 2\frac{\partial^2 u}{\partial D \partial d} = 0, \tag{4.26}$$

This is called the Clearbout poststack migration equation, forming the foundation for migration algorithms. As wave migration tends to make δ approach 0 infinitely, a migrated seismic section is the output of the wave migration process, which should meet the condition $D = d$, namely

$$u_{\text{mig}}(x,D,D) = u_{\text{mig}}(x,z,z) = u_{\text{mig}}(x,z) \tag{4.27}$$

Note that the data used in the prestack migration are different from the data in the poststack migration, so different migration equations should be used for the prestack migration, which will be discussed in Section 6.2. Many migration algorithms involve the integral solution of wave equations, which will be discussed in the following chapter.

The poststack wave migration processing produces stacked and migrated sections that show images of underground reflectors without extensive distortions. Figure 4.2 shows the migrated section generated by using the stacked section shown in Figure 4.1(a) as the input and the velocity model shown in Figure 4.1(b). The migration algorithm used is called depth migration, and it produces the migrated section shown in Figure 4.2. It is easy to find that most of the distortions that appeared in Figure 4.1(a) disappeared in the migrated section.

So far, most of the wave equations employed in reflection seismology, as well as their brief applications, have been studied. In summary, the wave equations employed in 2D seismic data processing are outlined in Table 4.1, together with the boundary conditions.

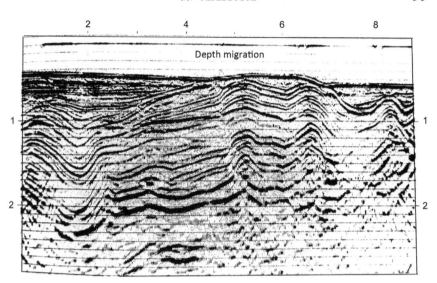

FIGURE 4.2 The migrated section generated by using the stacked section shown in Figure 4.1(a) as the input and the velocity model shown in Figure 4.1(b). *Source: James (1989).*

4.7. REFLECTORS

The target of reflective seismology, called underground reflectors, is the source of reflective waves. Generalized reflectors mean all the sources that produce waves appearing in the prestack seismic data, including scatter waves, multiples, and diffraction waves. Reflectors in a more restricted sense mean generators of the primary reflective waves, which exist in stacked seismic data without scattered and diffracted waves and are relevant to the soft boundaries defined in physical mathematics, which correspond to the source of reflecting waves. The reflectors discussed in physical mathematics belong to the restricted sense. In sedimentary basins, the reflectors defined by the soft boundaries correspond to sedimentary formations. In production seismology, the reflectors of the restricted sense enclosed also oil/gas reservoirs, as well as pores filled with fluid as discussed in Section 3.7. As mentioned in the last section, the migration process reveals the wave field directly created by reflectors, which are targets of seismic prospecting. Thus, we need to further discuss mathematical definitions of the reflectors, to improve the theoretical basis of reflection seismic processing and interpretation.

We have to physically classify the seismic source into two types in reflection seismology: the active source and the passive source. The physical excitation belongs to the active source, while the reflectors are the passive sources that correspond to the soft boundaries in mathematical physics. We will demonstrate in Chapter 5 that the boundary conditions

TABLE 4.1 Overview of Equations and Their Boundary Conditions Employed in 2D Seismic Processing

No	Processing Name	Input/Output Gathers	Equations Used	Variation Factors	Boundary Conditions
1	Statics	CSP	2D elastic wave equation	Source locations	Moving from surface relief to a base level beneath
2	Muting	CSP	2D elastic wave equation	Direct waves and head wave attenuation	Removing surface boundary
3	Deconvolution	CSP	1D acoustic equation	Source wavelet	
4	2D filtering	CSP	2D elastic wave equation	Attenuation of pure shear and surface waves	
5	Multiple suppression	CSP	1-2D acoustic equation	Multiples deduced	Removing the seabed boundary
6	NMO	CMP/zero-offset gather	2D acoustic equation	Transformation to 1D acoustic wave	Relocating receiver points
7	Poststack migration	Zero-offset gather/stacked and migrated section	2D migration equation	Backpropagation of primary reflective waves to approach reflectors	Imaging the soft boundaries

of an initial-boundary value problem of the wave equations can be moved to its source item, that is the soft boundaries are mathematically equivalent to some kind of source. We may denote interfaces between sedimentary formations as $\{\Gamma_i\}$, including all the soft boundaries (Section 3.2). The boundary conditions in Eqn (4.20) can be written as

$$u_{\Gamma_i}^+(x',t) = u_{\Gamma_i}^-, x' \in \{\Gamma_i\}, \quad i = 1, 2, ..., I \tag{4.20}$$

where the superscript "+" stands for the upper side of the ith boundary Γ_i and superscript "−" stands for the lower side, and I is the total number of boundaries. The boundary conditions Eqn (4.20) in the initial-boundary value problems can be transformed into the source items.

The passive source strength and polarity of an infinite flat interface are dependent on wave impedance difference of the media between the two sides of the interface, the interface shape, and the incident angle from the active source. The impedance difference between the two sides of the interface can be defined as reflectivity that is indicated by the ratio as (Sheriff, 1983, 1984; Telford et al., 1990)

$$R = \frac{\rho_2 V_2 - \rho_1 V_1}{\rho_2 V_2 + \rho_1 V_1}$$

where ρ is the density and V is the wave velocity, "1" denotes the upper rock layers and "2" the lower layer. To extend the flat reflectors to 2D cases, interfaces can be described as continuously relieved surfaces composed of points at $(x, z) \in \{\Gamma\}$, where interfaces are marked as the set $\{\Gamma\}$. The reflectivity function, as a spatial function to describe reflectivity, can be written as

$$R(x,z) = \gamma \frac{\rho_2(x,z)V_2(x,z) - \rho_1(x,z)V_1(x,z)}{\rho_2(x,z)V_2(x,z) + \rho_1(x,z)V_1(x,z)}, \quad (x,z) \in \{\Gamma\} \tag{4.28}$$

where γ is a parameter related to the boundary shape. Parameters related to boundary shape include curvature and dip. The reflectors have nonzero reflectivity $R(x,z)$ at the boundary Γ, which are equivalent to the passive sources.

We look at simple 1D problems first. To put the reflectors Eqn (4.20) into the source items of wave Eqn (4.19), we assume that $R(x,z)$ is looked upon as the primary sources that shoot simultaneously the wave field by the primary sources located on surface Γ is equivalent to real reflection wave field. Therefore, the source items in the acoustic equation can be written as the product of a wavelet function $W(t)$ and the reflectivity function by the following curve integral:

$$F(x,z,t) = W(t)u_0(x,z,t) \int_\Gamma \theta(x',z')R(x',z')ds \tag{4.29}$$

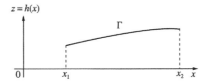

FIGURE 4.3   The interface function.

$$(x', z') \in \Gamma \qquad (4.30)$$

where u_0 is the incident wave field and ds is the arc differential of curve Γ. The reflection waves are excited by the incident waves $u_0(x,z,t)$; the weight function $\theta(x,z)$ is related to the angle between the incident wave and the normal direction of Γ.

Then, we consider 2D reflectors with no change outside the profile (x,z). Assume that interface Γ can be described as a continuous curve with respect to x as (Figure 4.3):

$$\Gamma : z = h(x) \qquad (4.31)$$

If this is called the interface function and h is the relative elevation, then the curve integral of the source term can be written as

$$F(x, z, t) = W(t)u_0(x, z, t)$$

$$\int_{X_1}^{X_2} \theta(x', z') \, R(x', z') \sqrt{1 + \left[\frac{dh(x')}{dx}\right]^2} \, dx', \quad (x', z') \in \Gamma \qquad (4.32)$$

It is assumed in Eqn (4.32) that the reflectivity function $R(x,z) = 0$ outside the boundary $(x, z) \in \Gamma$. So far, the weight function of the incident wave $u_0(x,z,t)$ and the angle $\theta(x,z,t)$ have not been given. The amplitude of the incident wave can be normalized after a data processing technique called amplitude equalization, so we can assume that $u_0(x,z,t) = 1$ after the equalization process.

A real sedimentary basin contains several formations, causing multiple interfaces corresponding to the boundary conditions in Eqn (4.20). In such cases, interfaces can be defined as a set of $\Gamma_i, (i = 1, 2, ..., m)$ as a set of functions,

$$\{\Gamma_i\} : z(x_s) = h_i(x_s), \quad i = 1, 2, ..., m$$

where h_i is the relative elevation and $\{\Gamma_i\}$ is a set of nonintersecting continuous curves in 2D cases. Then, the source term in its equivalent wave equation for simultaneously exciting reflectors after equalization can be simplified as

$$F(x,z,t) = W(t) \sum_{k=1}^{m} \int_{X_{1i}}^{X_{2i}} \theta_k(x_s,z_s) R_k(x_s,z_s) \sqrt{1 + [h'_k(x_s)]^2} dx_s, \quad (x_s,z_s)\{\Gamma_i\}$$

(4.33)

Equation (4.33) defines a simplified mathematical model for layered reflectors, useful to describe the relationship between reflection wave field u and the passive sources in a sedimentary basin, namely the interface boundary set $\{\Gamma_i\}$. One of the goals of reflection seismology is to locate the boundary set $\{\Gamma_i\}$ and to calculate its reflectivity function $R_i(x,z)$ accurately. Equation (4.33) can be put into the right source items of Eqn (4.22), obtaining the one-way wave equation that is excited by reflectors traveling in one direction to ground receivers as

$$\frac{\partial^2 u}{\partial x^2} + \frac{\partial^2 u}{\partial z^2} - \frac{1}{V^2(x,z)} \frac{\partial^2 u}{\partial t^2}$$
$$= -W(t) \sum_{k=1}^{m} \int_{-\infty}^{\infty} \theta_k(x_s,z_s) R_k(x_s,z_s) \sqrt{1 + [h'_k(x_s)]^2} dx_s,$$

(4.34)

where $V(x,z) = c(x,z)/2$, and $c(x,z)$ is the wave velocity, and

$$(x_s,z_s)\{\Gamma_k\} : z_k = h_k(x_s), \quad k = 1, \ldots, I$$

(4.35)

Equation (4.34) represents the waves simultaneously inspired by reflectors, which is treated as the physical foundation for 2D stacking and migration. The goal of seismic acquisition and data processing is to get the wave field $u(x,z,t)$ that satisfies Eqn (4.34); further quantitative seismic interpretation is to calculate $R_k(x_s,z_s)$ based on $u(x,z,t)$. If one assumes that $V(x,z) = c(x,z)$ in the above formula, the corresponding two-way wave equation can be easily obtained.

In practical calculations, the source items in the above two formulas can be simplified further in a discrete format. If one assumes that $x = i\Delta x$, where Δx is the trace interval, $i = 1, 2, 3, \ldots I$, and I is the number of gridding cells, the force source term can be expressed as

$$F(i,k,t) = W(t) \sum_{k=1}^{k} \sum_{i=1}^{I} \theta_k(i \cdot k) R_k(i,k) \sqrt{\Delta x^2 + \Delta h_k^2(i)}$$

(4.36)

Reflection seismology used in structure imaging in a sedimentary basin and localization of oil/gas reservoirs should be considered as an inter-disciplinary subject of wave theory and geological structure, among which the wave propagation study belongs to physics while that on geological formation location belongs to sedimentary geology. The research object in reflection seismology is to establish equations linking the wave theory and geological structures, which have the same locations as the reflectors simultaneously excited by active shots. Although the processing wave field is not purely real, it simply represents images equal to sedimentary formations. The so-called reflectors in sedimentary basins can be defined quantitatively by Eqns (3.33)–(3.66), providing a simplified tool for the reflection seismic processing and interpretation.

The continuous velocity model discussed in this section may make only little advance than the traditional layered models, but this progress helps a lot in mathematical calculation. The primary difference between continuous velocity model and homogeneous layered models lies in that the wave equations with constant coefficients should be used for oversimplified models while wave equations with variable coefficients can be used for the continuous model. The continuous model introduced in this section can be well easily applied to the finite difference method that is based on the concept of dividing studied media into finite units of small scales. The advantages of using this procedure are that the boundary conditions between adjacent units can be autosatisfied. A Fortran program for applying this procedure is presented in the Appendix, for the readers.

Mathematically, accurate earth models in reflection seismology and VSP must be defined by generalized functions, because discontinuities included in earth models do not satisfy wave equations, but only the mass and energy conservation laws. To understand how reflective signals are generated from underground velocity discontinuities, we have to further discuss descriptions of the seismic discontinuities that exist in wave propagation media. As discussed in Chapters 2 and 3, discontinuities of physical parameters, such as the stress or displacement components, mark the boundary conditions. Furthermore, crossing of the discontinuity surfaces appears actuarially in a sedimentary basin, the crossing is the so-called singular point of discontinuity in mathematics where the boundary conditions lose their meaning. Thus, a seismic velocity function $c(x,y,z)$ is not a function of common sense that has a one-to-one correspondence to its arguments, but it belongs to a kind of generalized function, which can be described mathematically only by the distribution theory. The intrinsic singular points included in the continuous velocity model can be divided into three types as follows:

1. Intersections of rock boundary surfaces where one has two contradictable boundary conditions.
2. Points at an angular unconformity where one cannot define the normal direction of the boundaries.
3. Terminations of the discontinuities, that is thinning-out points. The velocity function $c(x,y,z)$ varies continuously so that physical parameters may converge along the discontinuity surfaces and finally make the surfaces disappear.

The intersections, angular points, and terminals mentioned above physically produce diffraction waves that cause the wave field to be more complicated. The wave field near these singular points becomes difficult to calculate, so some special procedures should be considered. On the other hand, these singular points provide more wave signals that are useful to seismic interpretation.

4.8. SUMMARY

It had been thought that the goal of reflection seismology is to improve the resolution and the S/N ratio of seismic data, which is widely understood and accepted by geologists. But geophysicists must have a deep understanding of the variations of physical models in each step of seismic processing, that is the variations of wave equations and their boundary conditions involved. The sequence of seismic data processing simplifies physical models as well as wave equations. After the analysis of seismic data processing based on equations and boundary conditions, the goals of seismic processing can be expressed as follows:

1. The first goal of seismic data processing is to transform the seismic data satisfying 3D elastic wave equations into waves that can be approximately described using the 1D homogeneous acoustic equation, or waves that are excited by a point pulse source described by the δ function.
2. The second goal of seismic data processing is to reduce or to simplify the boundary conditions satisfied by seismic data, making the wave field sharply correspond to underground reflection interfaces. The stacked seismic data sets obtained under these two goals turn original data sets into zero-offset gathers, showing a wave field that is similar to the pure reflection acoustic waves excited by vertically incident plane waves.
3. The third goal of seismic data processing is to transform the pure reflection waves into nonpropagated reflection waves directly emitted from underground reflectors. The stacked and migrated waves are proportional to the reflective function of the reflectors, so the stacked and migrated sections represent the image of the reflectors.

Integral Solutions of the Wave Equation with Boundary and Initial Value Conditions

The previous four chapters discussed seismic wave propagation in the Earth and the representation and transformation of seismic reflection data. Seismic data processing has been explained in detail by some books (Geldart and Sheriff, 2004) and briefly introduced in the past chapter.

However, more aspects exist in theoretical reflection seismology. The numerical solution of the wave equations together with boundary and initial conditions can be found by using the finite element and finite difference methods. As this book is devoted to outline the theories in reflection seismology, it does not discuss these numerical solution methods in details (but if the book is used as a textbook, it is better to assign homework to students by using program included in the appendix for solving the wave equations with the finite difference method). Although the analytic solutions of the wave equation with boundary and initial conditions have few direct applications, they are very important in theoretical researches, especially for the inverse problems. After a systematic discussion of the wave differential equations, we will discuss the integral solutions of the wave equation with the corresponding boundary and initial conditions.

Finding analytic solutions of the wave equation with boundary and initial conditions is a part of direct problems, which is a subject of mathematical physics. In the early to middle twentieth century, mathematicians already achieved excellent research results on mathematical physics. In the twenty-first century, the purpose of analytic solutions of the wave equation is not improving the seismic data acquisition any more but mainly focuses on information retrieval and inversion of the seismic reflection data. The acoustic wave equation is a second-order linear hyperbolic partial differential equation. Its operator is called a second-order linear hyperbolic partial differential operator (more strictly speaking, it is a pseudo-differential operator). It is easy to guess that the reverse operator of a differential operator is an integral operator. So, the solution of a direct wave equation problem comes from solving a partial differential equation and the solution of an inverse wave equation problem comes from solving an integral equation, while the bridge between the two kind of problems belongs to the integral solution of a direct problem.

Table 5.1 shows some types of the integral solutions of wave equations with boundary and initial conditions. The so-called basic solution of the wave equation is not related with the boundary conditions, and it is a solution of the initial value problem in infinite homogeneous isotropic medium, which is already discussed in Section 1.4. The general solution is related to the general boundary conditions in homogeneous isotropic medium, while the particular solution is related to the particular boundary conditions in homogeneous isotropic medium. In Section 3.3, we have discussed the particular solution of elastic wave equation in horizontally layered medium. The special solutions of boundary value problems for hyperbolic wave equation have been discussed in details in many mathematic monographs, so they are not the main subjects of this chapter. On the other hand, the generalized solution refers to the solution of the initial value problem of wave equations with variable coefficients in

TABLE 5.1 Types of the Integral Solutions of Wave Equations with Boundary and Initial Conditions

Solution Type	Coefficient in the Equation	Propagation Medium	Boundary Conditions	Specified Boundary Value or Parameters
Basic solution	Constant coefficient	Homogeneous and isotropic	No boundary	
General solution	Constant coefficient	Homogeneous and isotropic	General boundary	Wave field or its derivatives
Particular solution	Constant coefficient	Homogeneous and isotropic	Particular boundary	Geometry parameters of the boundary
Generalized solution	Variable coefficient	Inhomogeneous medium	General boundary	Parameters about heterogeneity

inhomogeneous medium, and can be further divided into the general solution and particular solutions. In the first two sections of this chapter, we will discuss the general solution of the wave equation with boundary and initial conditions. Thereafter, we will discuss the generalized integral solutions in the form of Green's functions in the last four sections.

5.1. INTEGRAL SOLUTIONS FOR MIXED CAUCHY BOUNDARY VALUE PROBLEMS

Cauchy problem is for calculation of the displacement of a wave field $u(x,y,z,t)$ in homogeneous medium with no boundaries, where the displacement $u(x,y,z,t)$ satisfies the constant coefficient wave equation (Tikhonov et al., 1961; Pearson, 1974; Eringen and Suhubi, 1975)

$$\nabla^2 u - \frac{1}{c^2}\frac{\partial^2 u}{\partial t^2} = f \tag{5.1}$$

And when $t = 0$, the initial conditions are given by

$$u(x, y, z, 0) = \Phi(x, y, z) \tag{5.2a}$$

and

$$u_t(x, y, z, 0) = \Psi(x, y, z) \tag{5.2b}$$

where Φ is the initial wave field and Ψ is the time derivative of the initial wave field. Physically, the Cauchy problem shows how to derive the characteristics of wave propagation given only the initial wave field and its derivative at a specified time. The Cauchy problem implies a hypothesis that the distribution area for both the initial wave field Φ and its time derivative Ψ is very limited when $t = 0$ and that the wave field distribution area will enlarge with the increase in energy diffusion when $t > 0$. On the other hand, given the wave field record and its derivative at a specified time $t = 0$, whether the early wave propagation in $t < 0$ can be derived or not is a problem belonging to seismic wave migration that has been discussed in Chapter 4.

First of all, let us consider the case without sources, i.e. $f = 0$ in Eqn (5.1). In an infinite homogeneous medium with no boundaries, the solution for the Cauchy problem can be found in the *Handbook of Mathematics* directly (e.g. Carl E. Pearson, 1974), while in a medium with boundaries, we need to define the wave field on the boundaries, and this problem will be discussed in the next section. In physics, wave field u in the wave equation represents the displacement potential of elastic wave, which can be applied to both longitudinal and shear waves.

For one-dimensional homogeneous wave equation, the general solution of the Cauchy problem is the so-called d'Alembert's formula

$$u(x, t) = \frac{\Phi(x - at) + \Phi(x + at)}{2} + \frac{1}{2c} \int_{x-ct}^{x+ct} \Psi(x') dx' \tag{5.3}$$

where the first term on the right-hand side of the equation corresponds to the wave field propagation toward both positive and negative directions and the second term is the delay effect of the nonpulse wave.

In the case of three dimensions, the general solution of the Cauchy problem is called Poisson's formula (Tiknonov and Samarskii, 1963; Pearson, 1974)

$$u(x, y, z, t) = \frac{1}{4\pi c^2} \left[\frac{\partial}{\partial t} \iint_S \frac{\Phi}{r} ds + \iint_S \frac{\Psi}{r} ds \right] \tag{5.4}$$

where S is the surface of a sphere with radius r centered at point (x, y, z),

$$r^2 = (x - x')^2 + (y - y')^2 + (z - z')^2 = c^2 t^2$$

where (x', y', z') is a point on the sphere surface. Using the above equation, the Poisson's Formula (5.4) can be written as:

$$u(x, y, z, t) = \frac{1}{4\pi c} \left[\frac{\partial}{\partial t} \iint_S \frac{\Phi}{t} ds + \iint_S \frac{\Psi}{t} ds \right] \tag{5.5}$$

By applying Poisson's formula or d'Alembert's formula, the wave field propagation for $t > 0$ can be calculated by the given initial wave field and its time derivative.

In the case of two dimensions, let the displacement $u(x,z,t)$ satisfy the constant coefficient homogeneous wave equation

$$\frac{\partial^2 u}{\partial x^2} + \frac{\partial^2 u}{\partial z^2} + \frac{1}{c^2}\frac{\partial^2 u}{\partial t^2} = 0$$

And when $t = 0$, the initial conditions are given by

$$u(x,z,0) = \phi(x,z)$$
$$u_t(x,z,0) = \psi(x,z)$$

The solution of the Cauchy problem for an infinite homogeneous medium can be expressed by the two-dimensional (2D) Poisson's formula

$$u(x,y,z,t) = \frac{1}{2\pi c}\left[\frac{\partial}{\partial t}\iint_S \frac{\Phi dx'\,dz'}{\sqrt{(ct)^2 - (x-x')^2 - (z-z')^2}}\right.$$
$$\left. + \iint_S \frac{\Psi dx'\,dz'}{\sqrt{(ct)^2 - (x-x')^2 - (z-z')^2}}\right]$$

(5.6)

where S is the intersecting circle between a plane having $y = 0$ and a sphere of radius $r = ct$. The sphere is centered at point (x,z) and is defined by the formula

$$(x-x')^2 + (z-z')^2 = c^2 t^2$$

Now, let us consider the case with the source term, i.e. $f \neq 0$ in Eqn (5.1). When wave equation becomes inhomogeneous, the term $f(x',y',z',t')$ on the right-hand side of the equation corresponds to the excitation source of the wave field, where t' is called the delay time of the source wave field

$$t' = t - \frac{r}{c}$$

Thus a term of u_1 for the delay wave should be added to the solution for the Cauchy problem, which is written as:

$$U = u + u_1$$

where u is the integral solution expressed as in the Poisson Formula (5.5), and u_1 is the delay potential of the source wave field. The delay potential needs to be added to the solution U for the Cauchy problem in three-dimensional (3D) problems, which is

$$u_1(x,y,z,t) = \frac{1}{4\pi c}\iiint \frac{f(x',y',z',t')}{r}\,dx'\,dy'\,dz'$$

(5.7a)

In 2D problem, it is

$$u_1(x,z,t) = \frac{1}{2\pi c} \int_0^t \left\{ \iint_S \frac{f(x,z',t')}{\sqrt{(ct)^2 - (x-x')^2 - (z-z')^2}} dx' \, dz' \right\} dt' \quad (5.7b)$$

and the corresponding integral domain S is given by the circle

$$(x-x')^2 + (z-z')^2 = c^2(t-t')^2$$

In one-dimensional problems,

$$u_1(x,t) = \frac{1}{2c} \int_{x-c(t-t')}^{x+c(t-t')} f(x',t') dt' \quad (5.7c)$$

As mathematicians derived the integral solutions for the mixed Cauchy boundary value problems, Eqns (5.7a)–(5.7c) shows a clear physical meaning for interpreting wave propagation process. For instance, take the x as the vertical axis. Then, both Eqns (5.3) and (5.7c) imply the existence of upgoing and downgoing waves. It is very helpful for us to understand the reflection wave as a type of upgoing waves.

5.2. THE KIRCHHOFF INTEGRAL FORMULA FOR THE BOUNDARY VALUE WAVE EQUATION PROBLEMS

Poisson's formula shows wave propagation when only the initial wave field and its derivative in the infinite medium with no boundaries are given. Seismic reflection deals with the signals observed at the surface and reflected from the internal earth with no observation of the initial wave field; we need to use the Kirchhoff formula which will be discussed as follows. Kirchhoff formula can be used to calculate the wave propagating in homogeneous medium within a boundary from the wave field and its derivative on the border. For seismic reflections, the Kirchhoff formula can be derived from vector analysis and Green's integral theorem (see Tikhonov et al., 1961, Chapter 5; Pearson, 1974; Eringen and Suhubi, 1975), but based on assumptions of homogeneous medium and constant velocity c. Let x be a point in the area Ω bounded by S; the Kirchhoff formula can be written as:

$$u(x,t) = \iint_S \left\{ \frac{1}{r} \left[\frac{\partial u(x',t')}{\partial n} \right] - [u(x',t')] \frac{1}{\partial n} \left(\frac{1}{r} \right) + \frac{1}{cr} \left[\frac{\partial u(x',t')}{\partial r} \right] \frac{\partial r}{\partial n} \right\} dS_{x'}$$

$$+ \frac{1}{4\pi} \iiint_\Omega \frac{f(x',t')}{r} d\Omega_{x'} \quad (5.8a)$$

where u is the wave field satisfying the inhomogeneous wave Eqn (5.1); n is the unit outward normal vector, which represents the derivative direction; f is the source of force, presented on the right-hand side of the wave equation; the bracket [*] denotes a phase shift r/c in time; r is the distance from integral element x' to the point x (see Section 5.4); and the delay time equals

$$t' = t - \frac{r}{c}$$

The first term in Eqn (5.8a) represents the wave from the border, the second term represents the energy diffusion due to wave propagation, the third terms represents the time-shift for the nonpulse wave, and the forth term represents the wave field of the source. Let

$$g(x, x', t') = \frac{\delta(ct - r)}{4\pi cr},$$

then substituting the delay time t' to the integral Formula (5.8a) and assuming there is no force source in the area Ω, Eqn (5.8a) can be written as

$$u(x, t) = \frac{1}{4\pi} \int dt' \iint_S \left\{ g(x, x', t') \frac{\partial u(x', t')}{\partial n} - u(x', t') \frac{\partial g(x, x', t')}{\partial n} \right\} dS_{x'}$$

(5.8b)

We will see this formula again in Eqn (5.65). In the next section, we will prove that the function g in the Formula (5.8b) is the Green's function of the point source wave equation in homogenous half space. This formula has very important application to the prestack wave migration in reflective seismic data imaging, which will be discussed later in Chapter 6.

When the vibration can be represented by the superposition of harmonic vibrations, the wave equation in the frequency domain can be derived by using the time Fourier transform of wave Eqn (5.1), which represents the propagation of harmonic vibrations:

$$\nabla^2 u(x, \omega) + k^2 u(x, \omega) = -f(x', \omega); \quad k = \frac{\omega}{c} \qquad (5.9)$$

Letting the amplitude of the source harmonic vibration as f_0 and amplitude of the harmonic vibration of wave field u to be u_0, we have

$$[u] = u_0(x)e^{i(\omega t - kr)}; \quad [f] = f_0(x)e^{i(\omega t - kr)}$$

$$\left[\frac{\partial u}{\partial n}\right] = \frac{\partial u_0}{\partial n}(x)e^{i(\omega t - kr)}; \quad \frac{1}{c}\left[\frac{\partial u}{\partial r}\right] = iku_0 e^{i(\omega t - kr)}$$

On substituting the above four equations to Eqn (5.8), we have

$$u(x) = \frac{1}{4\pi} \iint_S \left\{ \frac{e^{-ikr}}{r} \frac{\partial u_0}{\partial n} - u_0 \frac{\partial}{\partial n} \left(\frac{e^{-ikr}}{r} \right) \right\} dS_{x'} + \frac{1}{4\pi} \iiint_\Omega f_0(x') \frac{e^{-ikr}}{r} d\Omega_{x'}$$

(5.10)

This is the Kirchhoff formula in the frequency domain in homogenous medium. Equation (5.10) is often used in reflection seismology when the vibration can be represented by the superposition of harmonic vibrations. The first term of Eqn (5.10) represents the wave from the border, the second term represents the energy diffusion with the wave propagation, and the third term represents the wave field of the force source. The time-shift effect from the nonpulse wave is already represented in the frequency spectrum, so Eqn (5.10) has less terms than Eqn (5.8a).

Now let us take an example for the application of Kirchhoff formula in reflection seismology, i.e. convolution model of zero-offset seismic traces (see Qian Rongjun, 2008). Suppose there is a reflective surface S of finite area; we want to compute the wave field in area Ω, which is on the same side of the surface S where the point source is located. Assume the point source produces harmonic wave with a single frequency ω. Receivers are located different from the source, the distance from the point source to surface element ds is r_0, and wave velocity in the homogenous medium is c. The reflective wave field on the surface element ds is:

$$[u] = \frac{A_0}{r_0} e^{-i\omega \left(t - \frac{r_0}{c} \right)}$$

where the reflective coefficient on surface S is a constant A_0. Given the wave field on the surface, we have to compute the normal derivative for using Kirchhoff formula. In zero-offset case, i.e. when $r = -r_0$, we have

$$\frac{\partial r_0}{\partial n} = -\frac{\partial r}{\partial n}$$

and

$$\left[\frac{\partial u}{\partial n} \right] = \frac{\partial r_0}{\partial n} \left(-\frac{1}{r_0} - \frac{i\omega}{c} \right) [u]$$

Substitute the above three equations into Eqn (5.8a) and using Kirchhoff Formula (5.10), it follows that

$$u(r,t) = \frac{1}{2\pi} \iint_S \left\{ -\frac{\partial}{\partial n} \left(\frac{1}{r} \right) + \frac{1}{cr} \frac{\partial r}{\partial n} \frac{\partial}{\partial t} \right\} \frac{A_0}{r} e^{i\omega \left(t - \frac{2r}{c} \right)} ds$$

(5.11)

where n is the unit outward normal vector of the surface S; the force term f in Eqn (5.8a) is zero in this equation. Because the reflective wave of zero-offset trace goes a round trip and the delay time of the direct reflective wave with respect to the source equals the round trip time

$$t' = t - \frac{2r}{c}$$

Let us derive the explicit form of the reflective wave assuming that the reflective surface S is a circular area on horizontal plane with radius d; the center of the circular surface is fit on the z-axis in a cylindrical coordinate system (x,θ,z), while S becomes plane $z = 0$; and ds is the symmetric ring with very small width dx. The zero-offset geophone is located above the surface S with a height h. The distance r from the zero-offset geophone to ds satisfies $r^2 = x^2 + h^2$. Because the surface element ds $= 2\pi dr$,

$$\frac{\partial r}{\partial n} = \frac{h}{r}, \quad \frac{\partial}{\partial n}\left(\frac{1}{r}\right) = -\frac{h}{r^3}$$

The above three equations show that the area integral in Eqn (5.11) can be computed by using the integral of a single variable r. Letting $q^2 = h^2 + d^2$, substitute the above three equations into Eqn (5.11); we have:

$$u(r,t) = A_0 \left[\int_h^q \frac{h}{r^3} e^{i\omega\left(t-\frac{2r}{c}\right)} dr + \int_h^q \frac{h}{cr^2}(i\omega)e^{i\omega\left(t-\frac{2r}{c}\right)} dr \right] \tag{5.12}$$

Using the subsection integral method, we have the reflective wave field as

$$u(r,t) = A_0 \left[\frac{1}{2h} e^{i\omega\left(t-\frac{2h}{c}\right)} + \frac{h}{2q^2} e^{i\omega\left(t-\frac{2q}{c}\right)} \right] \tag{5.13}$$

The Formula (5.13) suggests that there are two components in the zero-offset wave field reflected from a finite surface S in the 3D homogenous medium. The first component is the reflected wave that can be received only above the surface S and is represented by the first term on right-hand side of Formula (5.13). Its phase shift is $T_1 = 2h/c$, and the amplitude is inversely proportional to the distance h. The second component is the diffracted wave and is represented by the second term on right-hand side of Formula (5.13). Its phase shift is $T_2 = 2q/c$ and the amplitude is inversely proportional to square of the distance q. When the radius of the reflective surface S increases, the amplitude of the diffracted wave becomes smaller and phase shift becomes larger. The diffraction wave tends to 0 if the reflection surface S tends to infinity. In conclusion, the integral solution of Eqn (5.13) shows that the reflected wave field is composed of the reflected wave and diffracted wave. Reflected wave is excited from the reflective surface with its amplitude attenuating proportional to the distance, while diffraction wave is excited from the lateral boundary of the reflective surface with its amplitude attenuating proportional to the square of propagating distance.

Qian (2008) also proved that zero-offset traces can be further represented by the convolution integral model. Rewriting Eqn (5.12) as

$$u(r,t) = A_0 \int_h^q \left[\frac{h}{r^3} + \frac{h}{cr^2}(i\omega) \right] e^{i\omega\left(t - \frac{2r}{c}\right)} dr, \qquad (5.14)$$

and denoting the propagation time as

$$\tau = \frac{2r}{c}, \quad d\tau = \frac{2}{c} dr, \quad p = \frac{2h}{c}$$

Equation (5.14) can be rewritten as

$$u(r,t) = 2A_0 \int_h^\infty \left[\frac{p}{c\tau^3} + \frac{i\omega p}{c\tau^2} \right] e^{i\omega(t-\tau)} d\tau = \int_h^\infty R(\tau)w(t-\tau)d\tau \qquad (5.15)$$

where

$$R(\tau) = 2A_0 \left[\frac{p}{c\tau^3} + \frac{i\omega p}{c\tau^2} \right]$$

$$w(t-\tau) = e^{i\omega(t-\tau)}$$

Equation (5.15) is the convolution model of zero-offset traces obtained by using the wave equation in the frequency domain and Kirchhoff formula, meaning that a reflection seismic trace composed of the direct reflective wave and diffracted wave from reflective surfaces can be expressed as convolution of $R(t)$ and $w(t)$. $R(t)$ in the equation is equivalent to the amplitude function, and $w(t) = \exp(i\omega t)$ is equivalent to the phase shift function. When the vibration source can be treated as superposition of harmonic vibrations, the convolution model in Eqn (5.15) can be used to calculate zero-offset seismic trace.

The Kirchhoff formula is very important in seismic wave theory, especially for reflection seismology. As a summary of this section, the following three statements are listed, which are very helpful for understanding the seismic reflection waves.

1. Although solving integral solutions needs both boundary wave field and its normal derivative, the normal derivative of the boundary wave field can be obtained from the wave field conversion. Therefore, there are no unconquerable difficulties in calculating the wave propagation in homogeneous medium within a boundary.

2. In the stacked sections, the reflected wave and diffracted wave exist at the same time. Reflected wave comes from the reflectors that correspond to the soft boundary conditions. Diffracted wave comes from diffractors that correspond to the termination of the soft boundary conditions, called singularity in mathematics, or the boundary discontinuity (Sjöostrand, 1980). In the poststack migration section, the energy of reflected wave should be located at the position of the reflectors, while the energy of diffracted wave should be focused on the boundary discontinuity.
3. The interpretation of a seismic section requires two steps. The first step is accurate positioning of reflectors and diffractors, i.e. correct imaging. The second step involves quantitative estimation of the properties of reflectors and diffractors, such as seismic velocity inversion or estimation of reflectivity function.

5.3. THE GREEN'S FUNCTION OF BOUNDARY VALUE PROBLEMS FOR WAVE MOTION

5.3.1. The Green's Function Method

In the previous section, we have discussed the integral solution of wave equation with boundary and initial conditions. The general solution of wave equation with boundary and initial conditions in inhomogeneous medium will be discussed in this section. In applications, solving a complicated math—physics problem needs wisdom and skills, which belong to the domain of applied mathematics. Green's function method is a successful example developed in applied mathematics. It is also an important tool to solve partial differential equations with boundary and initial conditions, widely used in applied geophysics.

Let us take an ordinary differential equation as a simple example to illustrate the Green's function (Budak et al., 1962; Frittman, 1965). Let a differential operator L satisfy

$$Lu \equiv \frac{d^2 u(x)}{dx^2} = F \tag{5.16}$$

We know that the solution of Eqn (5.16) has an integral form that must depend on the complexity of the function of force F on the right side. Taking the simplest form of F, we have for single-point pulse

$$Lg = \frac{d^2 g(x, x')}{dx^2} = \delta(x, x') \tag{5.17}$$

where g is the Green's function of operator L, which is the integral kernel of inverse operator of L:

$$u = L^{-1}F = \int g(x, x')F(x')dx' \qquad (5.18)$$

Equation (5.18) can be proved as follows:

$$Lu \equiv \frac{d^2u(x)}{dx^2} = F = \int \delta(x - x')F(x')dx' = \int LgF(x')dx'$$
$$= L \int g(x - x')F(x')dx'.$$

Equation (5.18) shows that there are no associations between Green's function and the complexity of the source terms in a differential equation. In terms of the differential equations with complicated sources, a key to solving a boundary value problem can be to find the Green's function of the equation first, because the properties of the differential operator in the equation have already been embodied by the Green's function. After the Green's function is obtained, although there are infinite number of solutions for ordinary differential Eqn (5.17), particular solutions can be found by calculating integral Eqn (5.18) after given boundary conditions. In the example above, let the range of x be the interval (0,1) and let $u(x)$ to be a continuous function in the interval, which satisfies boundary conditions:

$$u(0) = u(1) = 0$$

A solution of the equation $Lu = F$ can be expressed by

$$u(x) = \int_0^1 g(x, x')F(x')dx'$$

This example shows that the solution of a differential equation with boundary and initial conditions can be expressed by integral forms as long as the Green's function is given. With Eqn (5.18) extended to the 3D space, the Green's function for a solution of second-order partial differential equations with boundary and initial conditions can be written as

$$u(x, t) = \int_V g(x, x')F(x')dx' \qquad (5.19)$$

where x' is the source point and arbitrary observation point $x \in V$, $x' \in S$; $x \neq x'$. It should be noted that the Green's function method cannot be used to zero-offset wave field records, because V is the volume enclosed by S that represents as the outer boundary. In general, Friedman proved that the inverse of partial differential operator L is an integral operator and its kernel is the Green's function G of the partial differential operator L, which is the solution to the equation $LG = \delta$ in an n-dimensional space. The zero initial condition for wave equation implies the theorem of Laplace transform. When the number of spatial variables is more than 1, the Green's function is no longer an ordinary function but becomes a generalized function.

The Green's function method is classical in applied mathematics and using the Formula (5.19) to solve partial differential equations with initial and boundary value problems. This method includes the following steps: (1) calculating the Green's function G based on partial differential operator L and establishing the integral equation by employing Eqn (5.19) and (2) then making the integral equation linear by some numerical procedures if they are nonlinear, such as iterative procedures for approximate solutions. As the Green's function method can be applied to the wave equation with variable coefficients, it is widely accepted in reflection seismology.

5.3.2. Green's Function for the Wave Equation with Zero Initial Value Problems

In order to understand the Green's function method, we need to know the pulse point source defined by the right-hand side term of the equation $LG = \delta$. The pulse point source is located at point x' and excited at the time t'. First of all, let us look at the simplest case: Green's function for the wave equation with zero initial values. The Green's function of one-dimensional wave equation with initial and boundary value problems satisfies

$$\frac{\partial^2 g}{\partial x^2} - \frac{1}{c^2}\frac{\partial^2 g}{\partial t^2} = -\delta(x, x')\delta(t - t') \tag{5.20}$$

The corresponding initial conditions are given by

$$g(x, 0) = 0, \quad \frac{\partial g(x, 0)}{\partial t} = 0 \tag{5.21}$$

The conditions (5.20) and (5.21) imply that when $t < 0$ and $t = 0$, there are no vibrations in the area being studied, the excitation happens at $t = t'$, and the Green's function for the initial and boundary value problem is (Budak et al., 1962; Frittman, 1965)

$$g(x, x', t, t') = \frac{1}{2c}H[(t - t') - |x - x'|] \tag{5.22}$$

Substituting Eqn (5.22) into Eqn (5.20), we can prove the above formula where H is the step function,

$$H(t) = \begin{cases} 0, \ t < 0 \\ 1, \ t > 0 \end{cases}$$

Now let us compare it with the wave field excited by a unit source; the wave field also satisfies one-dimensional wave Eqn (5.20)

$$\frac{\partial^2 u}{\partial x^2} - \frac{1}{c^2}\frac{\partial^2 u}{\partial t^2} = -\delta(x, x')$$

The general solution for the above equation is

$$u(x, t, x', t') = \begin{cases} 0, \ |x - x'| > c(t - t') \\ \dfrac{1}{2c}, \ |x - x'| \leq c(t - t') \end{cases} \tag{5.23a}$$

Comparing Eqns (5.22) and (5.23a), it is easy to find that the Green's function of one-dimensional wave equation with the zero initial value is the general solution for the one-dimensional wave equation.

Now let us compare the Green's function with a skillful case in which the force term can be transferred to the initial conditions. The one-dimensional wave equation can be written as homogeneous equation,

$$\frac{\partial^2 u}{\partial x^2} - \frac{1}{c^2}\frac{\partial^2 u}{\partial t^2} = 0$$

If the initial conditions are written as

$$u(x, 0) = 0, \frac{\partial u(x, 0)}{\partial t} = \delta(x, x')$$

The general solution for the mixed initial value Cauchy problem is

$$u(x, t, x') = \begin{cases} 0, \ |x - x'| > ct \\ \dfrac{1}{2c}, \ |x - x'| \leq ct \end{cases} \tag{5.23b}$$

which corresponds to the case when the pulse is excited at $t' = 0$ in Eqn (5.18). The example shows that when the source is the time pulse, the force term in the wave equation can be eliminated and be transferred to the initial conditions, if the source is assumed to be excited at $t = 0$. The force term and initial boundary value conditions are interchangeable for the wave equation with initial and boundary value problems, which will often simplify the process of deriving the integral solutions.

The Green's function of 3D wave equation with initial and boundary value problems satisfies

$$\nabla^2 g - \frac{1}{c^2}\frac{\partial^2 g}{\partial t^2} = -\delta(x, x')\delta(y - y')\delta(z - z')\delta(t - t') \tag{5.24}$$

We may temporally set the wave velocity in the coefficients to unity, namely, $c = 1$, according to the mathematician's "refining" preferences. The initial condition at the excitation moment $t' = 0$ is given by

$$u(x,y,z,0) = 0, \frac{\partial u(x,y,z,0)}{\partial t} = 0 \tag{5.25}$$

The solution of the initial and boundary value problems for both Eqns (5.24) and (5.25) can be obtained in the frequency domain after 3D Fourier transform. Using the wave vector $\mathbf{k} = (k_1,k_2,k_3)$, vector $x = (x,y,z)$ and vector $x' = (x',y',z')$, the norm of the wave vector is

$$k^2 = k_1^2 + k_2^2 + k_3^2$$

The Green's function can be expressed as a step function as

$$g(\mathbf{x},\mathbf{x}',t,t') = \frac{1}{4\pi^2|\mathbf{x}-\mathbf{x}'|} \int\limits_0^\infty \{\cos[k(t-t') - |\mathbf{x}-\mathbf{x}'|] - \cos[k(t-t')$$

$$+ |\mathbf{x}-\mathbf{x}'|]\}H(t-t')dk \tag{5.26}$$

A famous physicist, Dirac, has shown that the Green's function in Eqn (5.26) is actually a generalized function of a distribution

$$g(\mathbf{x},\mathbf{x}',t,t') = \frac{H(t-t')}{4\pi^2|\mathbf{x}-\mathbf{x}'|}\delta(t-t'-|\mathbf{x}-\mathbf{x}'|) - \delta(t-t'+|\mathbf{x}-\mathbf{x}'|) \tag{5.27a}$$

In other cases when wave velocity is not unity, x should be changed to x/c and x' changed to x'/c in the δ function of Eqn (5.27a).

When the vibration source is an excitation pulse at $t = 0$, the force term of the wave equation is transferred to the initial condition Eqn (5.25) and the basic solution for the Cauchy problem of 3D wave equations becomes

$$u(\mathbf{x},\mathbf{x}',t) = \frac{1}{4\pi c^2|\mathbf{x}-\mathbf{x}'|}\delta(ct-|\mathbf{x}-\mathbf{x}'|) \tag{5.27b}$$

which corresponds to the Green's function (5.24). So the Green's function is also a basic solution for mixed initial value Cauchy problems.

In the 2D space, when the initial conditions Eqn (5.21) are put into the Poisson Formula (5.6), the basic solution of the Cauchy problem becomes

$$u(x,y,x',y',t) = \begin{cases} 0, & r > ct \\ \dfrac{1}{2\pi c\sqrt{c^2t^2 - r^2}}, & r \le ct \end{cases}$$

where

$$r = \sqrt{(x-x')^2 + (y-y')^2}$$

and the corresponding Green's function is

$$g(x,y,x',y,t) = \begin{cases} 0, & r > c(t-t') \\ \dfrac{1}{2\pi c \sqrt{c^2(t-t')^2 - r^2}}, & r \le c(t-t') \end{cases}$$

Now, we obtained the analytical expressions of the Green's function in the one-dimensional, 2D, and 3D spaces in a homogeneous medium. Through the discussion on Green's function, we can understand deeply the properties of the Green's function method. However, the simple assumption of uniform medium is not good enough to deal with practical wave problems. In the next section, we will consider more complicated medium models and will discuss the applications of the Green's function.

5.3.3. Green's Function of the Wave Equation in Half Space with a Point Source

During excitation and recording of reflection seismic acquisition that occur on the ground, waves propagated in the upper half space can be ignored. So, it is necessary to study the Green's function for wave propagation in the half space. Take a look on the problem of wave equation in a half space with a pulse point source (Budak et al., 1962; Frittman, 1965; Pearson, 1974; Eringen and Suhubi, 1975). First of all, we assume the infinite isotropic homogeneous medium and zero initial value conditions, and also constant wave velocity in the medium. The wave field displacement components $u(x,t)$ satisfies

$$\nabla^2 u(\mathbf{x},t) - \frac{1}{c^2}\frac{\partial^2 u}{\partial t^2} = -\delta(\mathbf{x} - \mathbf{x}_s) \tag{5.28}$$

where $\mathbf{x}_s$ is position of the point source. The zero initial conditions present

$$u(\mathbf{x},t) = 0$$
$$u'_t(\mathbf{x},t) = 0, t \le 0$$

A general solution to this problem has been given in Sections (3.3) and (3.5) in the third chapter, i.e.

$$u(\mathbf{x},t) = \frac{1}{4\pi c^2 |\mathbf{x} - \mathbf{x}_s|}\delta\left(t - \frac{|\mathbf{x} - \mathbf{x}_s|}{c}\right) \tag{5.29a}$$

This formula shows that, in this point source problem, wave pulse moves forward with time only and the amplitude attenuates geometrically with the distance from the source. In the general case of nonzero initial conditions, namely, $t - t' > 0$, the time of the pulse generated is denoted by t' and the Green's function g satisfies

$$g(\mathbf{x}, \mathbf{x}', t, t') = \frac{1}{4\pi c^2 |\mathbf{x} - \mathbf{x}'|} \delta\left(t - t' - \frac{|\mathbf{x} - \mathbf{x}'|}{c}\right) \tag{5.29b}$$

The half space means that there are boundary conditions on the ground surface:

$$\frac{\partial u}{\partial z} = 0, \quad \text{for } z = 0 \tag{5.29c}$$

By ignoring the wave field in the upper half space, we can make even extrapolation of the lower half space to the upper half space, while the boundary condition (5.29c) is automatically satisfied. Even extrapolation is done by letting

$$u(x, y, -z, t) = u(x, y, z, t).$$

Notice that although the integral solutions in Eqns (5.3)–(5.27) of the point source boundary value problem can be applied by even extrapolation in the assumed whole space to give a good solution for the wave problem of the half space, this solution does not include any surface wave phases. Only after filtering the surface waves, the reflection seismogram can fit as the wave field provided by the even extrapolation solution, which one must consider in the seismic processing sequence.

Now let us discuss the wave equation with pulse point source in the frequency domain. Assuming the zero initial condition and constant acoustic velocity c in the medium, the Green's function G in the frequency domain corresponds to acoustic wave Eqn (5.28) and satisfies

$$\nabla^2 G(\mathbf{x}_s, \mathbf{x}, \omega) + \frac{\omega^2}{c^2} G(\mathbf{x}_s, \mathbf{x}, \omega) = -\delta(\mathbf{x} - \mathbf{x}_s)$$

It is easy to see that the Green's function G has been represented by Eqn (5.29a) in the frequency domain, which can be written as

$$G(\mathbf{x}_s, \mathbf{x}, \omega) = \frac{1}{4\pi |\mathbf{x}_s - \mathbf{x}|} e^{i\omega \frac{|\mathbf{x}_s - \mathbf{x}|}{c}} \tag{5.29d}$$

Performing even extrapolation of the upper half space, the boundary condition (5.29c) will be automatically satisfied. However, the assumption that the seismic wave energy will spread in the upper half space seems still different from the actual case. If the detector is on the ground, namely, $x = (x, 0, 0)$, using image-source method to solve the same acoustic wave equation with initial and boundary value problems, we have

$$G(\mathbf{x}_s, \mathbf{x}, \omega) = \frac{1}{2\pi |\mathbf{x}_s - \mathbf{x}|} e^{i\omega \frac{|\mathbf{x}_s - \mathbf{x}|}{c}} \tag{5.30}$$

Compared to Eqn (5.29b), the amplitude is increased by twice due to introduction of the surface boundary condition, which makes the seismic wave energy no longer diffusing to the upper half space. The expression (5.30) is accepted in reflection seismology and will also be used in seismic inverse scattering (see Chapter 7).

5.4. THE GREEN'S FUNCTION IN MEDIUM WITH LINEAR VELOCITY

As the first step of complicating the medium model, we will discuss the Green's function in the medium with linearly increased velocity model, which is a simplified model often used in reflection seismology. Huang Lianjie and the author (1991) have given analytic solutions for studying this problem. Suppose that in a Cartesian coordinate system the acoustic wave velocity increases linearly with depth

$$c(z) = a + bz, b > 0 \tag{5.31}$$

where a is the wave velocity on the ground, b is the velocity gradient underground, and $x = (x, y, z)$ is an arbitrary point underground. The wave equation in frequency domain corresponding to harmonic vibration is represented as the acoustic wave equation containing variable coefficients $c(z)$. The point source is located at $x' = (x', y', z')$ and the wave field u satisfies

$$\left(\nabla^2 + \frac{\omega^2}{c^2(z)}\right) u(x, x', \omega) = -\delta(x - x') \tag{5.32}$$

where the left-hand side of the equation is no longer a partial differential operator but a pseudo-differential operator.

Because the velocity is not constant, coordinate transformation is necessary. We introduce a new special vector τ whose three components are pseudo-propagation times along x-, y- and z directions as follows:

$$\tau_1 = \frac{x}{c(z)}$$

$$\tau_2 = \frac{y}{c(z)}$$

$$\tau_3 = \int_0^z \frac{1}{c(z')} dz' \tag{5.33}$$

The new coordinate system based on $\tau = (\tau_1, \tau_2, \tau_3)$ is called the pseudo-time coordinate system. By substituting Eqn (5.33) in Eqn (5.31), we have

$$\tau_3 = \frac{1}{b} \ln\left(\frac{a + bz}{a}\right) = \frac{1}{b} \ln\left(\frac{c(z)}{a}\right) \tag{5.34}$$

$$c(\tau_3) = ae^{b\tau_3} \tag{5.35}$$

The velocity varies exponentially with z in the new coordinate system. The acoustic wave Eqn (5.32) in the new coordinate system becomes

$$\left(\nabla^2 - b\frac{\partial}{\partial \tau_3} + \omega^2\right)u(\tau, \tau', \omega) = -\frac{1}{a}\delta(\tau - \tau') \tag{5.36a}$$

According to the definitions, the differential operators satisfied by the Green's function equation is the same as the wave field with a point pulse source but the source terms are different. Therefore, in the new pseudo-time coordinate system, the Green's function of the medium with linear velocity function is:

$$\left(\nabla^2 - b\frac{\partial}{\partial \tau_3} + \omega^2\right)G(\tau, \tau', \omega) = -\delta(\tau - \tau') \tag{5.36b}$$

By comparing the above two equations, we have

$$u(\tau, \tau', \omega) = \frac{1}{a}G(\tau, \tau', \omega) \tag{5.37}$$

Comparing the source terms on the right-hand side of the two Eqns (5.36a) and (5.36b), it can be seen that, unlike in a homogeneous medium, wave field u and its Green's function G are proportional to each other in the pseudo-time coordinate system and the proportion coefficient is $(1/a)$.

The Eqn (5.36b) satisfied by the Green's function G has an analytical solution. Here, we give the solution to the Eqn (5.36b) that will be proved later:

$$G(\tau, \tau', \omega) = \frac{e^{\frac{b\tau_3}{2}}}{4\pi|\tau - \tau'|}e^{\mp\sqrt{\omega^2 - \left(\frac{b}{2}\right)^2}|\tau - \tau'|} \tag{5.38}$$

When the vertical gradient of the velocity is small, i.e.

$$\frac{b}{2} \ll \omega \quad \text{then} \quad \sqrt{\omega^2 - \left(\frac{b}{2}\right)^2} \approx |\omega| \tag{5.39}$$

The high-frequency approximation of the Green function G represented by Eqn (5.38) can be written as

$$G(\tau, \tau', \omega) = \frac{e^{\frac{b\tau_3}{2}}}{4\pi|\tau - \tau'|}e^{i\omega|\tau - \tau'|} \tag{5.40}$$

The above equation suggests that, in the pseudo-time coordinate system, the phase shift of the high-frequency wave in the medium with linear velocity function and the phase shift in the homogenous medium are basically the same, while the amplitude must change by multiplying a factor related with the vertical gradient of the velocity, besides the geometric diffusion.

Now let us see the process of deriving the Eqn (5.38). By moving the origin to τ', i.e. letting $\tau' = 0$, Eqn (5.36b) becomes

$$\left(\nabla^2 - b\frac{\partial}{\partial \tau_3} + \omega^2\right)G(\tau, 0, \omega) = -\delta(\tau_1)\delta(\tau_2)\delta(\tau_3) \tag{5.41}$$

Applying Fourier transform on the first two components of τ, we have

$$\left(\frac{\partial^2}{\partial \tau_3^2} - b\frac{\partial}{\partial \tau_3} + \left(\omega^2 - k_1^2 - k_2^2\right)\right)G(k_1, k_2, \tau_3, 0, \omega) = -\delta(\tau_3) \tag{5.42}$$

where the Fourier transform is

$$G(k_1, k_2, \tau_3, 0, \omega) = \iint G(\tau_1, \tau_2, \tau_3, 0, \omega)e^{-i(k_1\tau_1 + k_2\tau_2)}d\tau_1 d\tau_2 \tag{5.43}$$

For the homogeneous equation of (5.42), its differential operator is not changed,

$$\left(\frac{\partial^2}{\partial \tau_3^2} - b\frac{\partial}{\partial \tau_3} + \omega_0^2\right)G(k_1, k_2, \tau_3, 0, \omega) = 0 \tag{5.44}$$

where

$$\omega_0^2 = \omega^2 - k_1^2 - k_2^2 \tag{5.45}$$

Because Eqn (5.44) is a second-order ordinary differential equation, its characteristic equation is

$$\lambda^2 - b\lambda + \omega_0^2 = 0,$$

which has two characteristic solutions

$$\lambda_{1,2} = \frac{b}{2} \mp i\frac{\sqrt{4\omega_0^2 - b^2}}{2} \tag{5.46}$$

Thus, one of the characteristic solutions of Eqn (5.44) must take the following form:

$$G(k_1, k_2, \tau_3, 0, \omega)e^{-i\omega_0 t} = Ae^{\frac{b\tau_3}{z}}e^{-i\omega_0\left(t - \frac{\sqrt{4\omega_0^2 - b^2}}{z\omega_0}|\tau_3|sign(\omega_0)\right)} \tag{5.47}$$

where A is the coefficient to be determined. By eliminating excess terms, we have

$$G(k_1, k_2, \tau_3, 0, \omega) = A e^{\frac{b\tau_3}{z} + isign(\omega_0) - \frac{\sqrt{4\omega_0^2 - b^2}}{z}|\tau_3|} \tag{5.48}$$

In the next step, we are going to substitute Eqn (5.48) into Eqn (5.42) and compute the integral of τ_3 on both sides. Using the condition that the integration of the δ function on the right side of the equation is 1, the coefficient to be determined is:

$$A = \frac{1}{\sqrt{4\omega_0^2 - sign(\omega_0)b^2}} \tag{5.49}$$

Substituting it into Eqn (5.48), the Green's function for the medium of linear velocity can be derived. After performing 2D inverse Fourier transform on it, we finally have

$$G(\tau, 0, \omega) = \frac{e^{\frac{b\tau_3}{2}}}{4\pi\sqrt{\tau_1^2 + \tau_2^2 + \tau_3^2}} e^{\mp i\sqrt{\omega^2 - \left(\frac{b}{2}\right)^2}\sqrt{\tau_1^2 + \tau_2^2 + \tau_3^2}} \tag{5.50}$$

After moving the origin back to an arbitrary position, i.e. substitute $0 = (0,0,0)$ by $\tau' \neq 0$, we then complete the proof of Eqn (5.38).

In short, there are three steps in calculating the Green's function in medium with linear velocity function: first, according to the velocity coefficients a and b, transform the Cartesian time coordinate system into the pseudo-time coordinate system by using Eqn (5.33); then, derive the Green's function by using Eqn (5.38); and finally, transform the Green's function acting in the pseudo-time coordinate system back to the common Cartesian coordinate system and derive the Green's function acting in the Cartesian time coordinate system by interpolation.

5.5. THE EIKONAL EQUATION AND THE TRANSPORT EQUATIONS

For Earth exploration, studying the medium with linear velocity function is not enough, and we must also examine the medium that can be described with arbitrary velocity functions. It is very difficult, even impossible, to represent the Green's function of the arbitrary medium in analytical expressions. So approximation models and the so-called WKBJ approximation method must be used (Aki and Richards, 1980; Bender and Orszag, 1988). After the approximation, the wave equation with variable coefficients is reduced to an eikonal equation and a transport equation, which are both first-order partial differential equations. This section will discuss the derivation and the application conditions of WKBJ method together with the eikonal equation and transport equation.

If assume the velocity $c(x)$ of the medium to be a smoothly differentiable function, where x denotes a point in 3D space, the harmonic wave equation (pseudo-differential equation) in the frequency domain can be written as

$$\left(\nabla^2 + \frac{\omega^2}{c^2(x)}\right)u(x, x', \omega) = -\delta(x - x') \tag{5.51}$$

From the discussion in the previous sections, it is known that if the wave velocity c is constant in the medium, the basic solution of the wave field is:

$$u(x, x', \omega) = Ae^{\frac{i\omega r}{c}} \tag{5.52}$$

where $r = |x - x'|$ is the distance between the source and a receiver. We can prove by substituting Eqn (5.52) into Eqn (5.51) that Eqn (5.52) is the basic solution of the wave field when c is constant. Equation (5.52) shows that the wave field is composed of two parts, i.e. amplitude function and phase function, which are not coupled. When the wave velocity varies slowly in the medium, $c(x)$ is not constant any more and Eqn (5.51) cannot be transformed into algebraic equations in the frequency domain by Fourier transform. In order to derive a simple solution for the direct problem of wave propagation in the velocity varying media, we can assume the wave field to be still composed of noncoupled two parts, i.e. the amplitude function and phase function. However, although the amplitude function can be separated from the phase function, it must relate with the frequency, and the general trend is that it attenuates with the increase of frequency; we must consider this fact in the approximate solutions. Another problem is related with r in Eqn (5.52). When the wave velocity $c(x)$ changes in the medium, the ray is no longer a straight line and r should be represented by a curved length, as ray's bending occurred. Let τ be a function that replaces a curved length r, and divide r by many ray segments based on the wave velocity $c(x)$. By applying these conventions, the approximate solution of the Eqn (5.51) can be expressed as:

$$u(r, \omega) = \omega^\beta e^{i\omega\tau(r)} \sum_{j=0}^{\infty} \frac{A_j(r)}{(i\omega)^j} \tag{5.53}$$

where τ becomes the propagation time and β is the phase correction coefficient. Most people ignore the phase correction and let $\beta = 0$. When the wave velocity variation rate is not slow, it is suggested to derive β by adding constraints according to the actual medium model, to further improve the approximation computation.

The approximate solution of the acoustic wave Eqn (5.53) is called WKBJ approximate solution, which has been proposed separately by G. Wentzel, H. Kramers, L. Brillouin, and H. Jeffreys in different research problems. When the frequency is very high, rays become nearly straight lines, τ can be simplified as the distance r between the two points, and the WKBJ approximate solution reduces to the geometrical optics approximate solution where the light goes always along a straight line.

The realization of the above ideas needs theoretical proof. Substituting the approximate solution (5.53) to wave Eqn (5.51), we have

$$\sum_{i=0}^{\infty}\left\{(i\omega)^{2-j}A_j\left[(\nabla\tau)^2 - \frac{1}{c^2}\right] + (i\omega)^{1-j}\left[2\nabla\tau\cdot\nabla A_j + A_j\nabla^2\tau\right] + (i\omega)^{-j}\nabla^2 A_j\right\} = 0$$

(5.54)

Because Eqn (5.51) contains the second-order partial differential of t, where the term ω has the highest order of 2. To decouple the amplitude function and the phase function, on the left side of Eqn (5.54), set the terms containing ω^2 with no-zero A_0 to zero. To begin with, letting the first term to be zero produces

$$(\nabla\tau)^2 = \left(\frac{\partial\tau}{\partial x}\right)^2 + \left(\frac{\partial\tau}{\partial y}\right)^2 + \left(\frac{\partial\tau}{\partial z}\right)^2 = N^2(x,y,z)$$

(5.55)

which is called eikonal equation of the propagation time $\tau(x,x')$, where

$$N^2(x) = \frac{1}{c^2(x)}$$

(5.56)

is called refractive index, which represents ray's bending and has nothing to do with the amplitude function. Substituting Eqn (5.56) into Eqn (5.54) and letting the sum of the terms with coefficients of the same ω powers to be zero, we have

$$A_0\nabla^2\tau + 2\nabla A_0\cdot\nabla\tau = 0$$

(5.57)

and

$$2\nabla\tau\cdot\nabla A_j\nabla^2\tau = -\nabla^2 A_{j-1} \quad j = 1, 2, \dots$$

(5.58)

These two formulas are called the transport equations, satisfied by the amplitude function $A(x,x')$. By using Eqn (5.57), people can obtain the first-order amplitude function solution $A_0(x,x')$ but forget the higher order approximations. Thus Eqn (5.57) is often called the transport equation only. In the case of higher order approximations, Eqn (5.58) must be applied. After solving Eqns (5.55)–(5.58), one can use Eqn (5.53) to get the approximation for the wave field.

The wave field Eqn (5.53), represented by the amplitude function and phase function, can also be used to represent approximate Green's functions. In the case of point source, suppose that the pulse point source is located at point x' and the wave velocity $c(x)$ in the medium is a smooth differentiable function of the second order, the wave equation representing the propagation of harmonic vibration in the frequency domain can be written as

$$\left(\nabla^2 + \frac{\omega^2}{c^2(x)}\right)u(x, x', \omega) = -\delta(x - x') \tag{5.59}$$

The corresponding Green's function satisfies

$$(\nabla^2 + k^2)G(x, x', \omega) = -\delta(x - x'); \quad k^2 = \frac{\omega^2}{c^2(x)} = \omega^2 N(x) \tag{5.60}$$

In the case of WKBJ approximation, if $c(x)$ changes in a scale that is much larger than seismic wavelength, then we can take first-order $(j = 0)$ solution in Eqn (5.53). The wave field u and the Green's function G can be written as:

$$u(x, x', \omega) = qA(x, x')e^{i\omega\tau(x,x')} \tag{5.61}$$

$$G(x, x', \omega) = qA(x, x')e^{i\omega\tau(x,x')} \tag{5.62}$$

where $\tau(x,x')$ is the propagation time, which depends on the velocity of $c(x)$ and q is a coefficient that depends on the dimension. Take $q = 1$ in the 3D case, and $q = (-i)1/2$ in the 2D case. Substituting Formulas (5.61) and (5.62) into Formulas (5.59) and (5.60), the problem of solving the Green's function becomes the problem of solving eikonal equation and transport Eqns (5.55)–(5.57). Of course, to solve practical problems, we must give the initial and boundary conditions as well, such as the location of the point source and the main direction of ray path.

From the eikonal function (5.55), it is easy to see that the magnitude of the wave slowness is

$$N = \frac{1}{c(x)} = |\nabla\tau| \tag{5.63}$$

The index N, which equals the wave slowness vector, can be written as $N = \nabla\tau$. Taking the minimum magnitude of the wave slowness, we have the Fermat's principle that says: the ray traveling along the path between any two points must have the shortest propagation time.

$$|\tau(x, x')| = \int_{x'}^{x} N(x'')dx'' = \min \tag{5.64}$$

Solving the eikonal equation to find the propagation time $\tau(x,x')$ is called ray tracing. There are many methods available for ray tracing; for the most commonly used ones, one may refer to the following authors' book: Hubral et al., 1992; Yang, 1997; Cerveny, 2001; Wu Rushan and Maupin, 2007.

The approximation of wave equation by the eikonal equation and transport equation is conditional, that is, the frequency must be high enough and higher the frequency, the more accurate is the approximation. There is the so-called asymptotic ray theory commonly used in seismic data processing, but it can make waveform distortions in the synthetic seismogram. Both ray tracing or solving the eikonal differential equation can be used as the numerical methods to produce the propagation time function $\tau(x,x')$, but substituting $N = \nabla\tau$ into the transport Eqn (5.57) for $A(x,x')$, one may find wave field divergence at some strange locations. These points of wave divergence are known as wave field singular points, where the WKBJ approximation is not valid. Therefore, the success of any method based on WKBJ algorithm depends on a careful search of the singular points, and then the Maslov asymptotic theory can be useful to improve the approximate solutions (Maslov, 1965; Sjöostrand, 1980), especially to deal with the singularity of the wave field.

5.6. THE SECOND-TYPE GREEN'S FUNCTION WITH NONHOMOGENEOUS BOUNDARY CONDITIONS

An equation in mathematical physics belongs to mathematical description of physical phenomena, because description is usually not unique and there are many equivalent descriptions. In Section 5.1, we have seen the case that the force term in the differential wave equation can be moved into the initial conditions, which means that a homogeneous equation with nonhomogeneous initial conditions is equivalent to a corresponding nonhomogeneous equation with homogeneous initial conditions. We can imagine that one may move the force term of the differential equation into the boundary conditions as well, namely, the homogeneous equation with nonhomogeneous boundary conditions may be equivalent to the nonhomogeneous equation with homogeneous boundary conditions. We will discuss this matter as follows. The wave propagation problems do not take a unique form, together with different initial or boundary conditions.

When we discussed the initial and boundary conditions of the Green's function for the wave equation in the previous section, the studied area is assumed to be infinite or the wave field is assumed to be zero on boundaries. In the following discussion, we will study further the second type Green's function for the wave equation, which is the Green's function with nonzero wave field boundary, but corresponding to homogeneous wave equation (Futterman, 1962; Tiknonov and Samarskii, 1963; Budak et al., 1964). Let a second-order partial differential operator be L, then the wave field function u satisfies the homogeneous equation with nonhomogeneous boundary conditions as

$$Lu(x) = 0 \quad (x \in V)$$
$$Bu(x) = b(x) \quad (x \in S) \tag{5.65}$$

where B is the operator acting on the boundary, $b(x)$ is the given wave field on the specified boundary S, and V is the volume enclosed by S. The simplest case is the unit operator, namely, $Bu = u$. If L is a second-order hyperbolic partial differential operator, Eqn (5.65) is called the wave equation with nonhomogeneous boundary conditions. Applying the Green's ideal, for an integrable function $b(x)$ on the surface of S, an integral solution of Eqn (5.65) can be:

$$u(x) = \int_S G_s(x, x')b(x')ds(x') \tag{5.66}$$

where G_s is called the second type Green's function, which satisfies

$$BG_s(x, x') = 0 \quad (x \in V, x' \in S, x \neq x') \tag{5.67}$$

and

$$\int_S BG_S(x, x')ds(x') = 1 \tag{5.68}$$

where V is the volume within the boundary S. The second type Green's function can be an ordinary function or sometimes a generalized function.

The second type of the Green's function has important applications. Wave equations in the frequency domain satisfy the conditions of the second type Green function, so the second type Green's function is widely used in applied seismology. It shows that nonhomogeneous equations or boundary value problems in practice can be solved by the superposition of a volume integral Eqn (5.18) and a curved surface integral Eqn (5.66). That is to say, we can derive the second type Green's function G_s from Eqn (5.62) by using recording data as the boundary conditions, find the transformation relationship between G_s and the Green's function G, and finally use Eqn (5.18) to derive the force term or the coefficient term in the nonhomogeneous equation.

To study the relationship between the second type Green's function G_s of the wave equation in the frequency domain and the ordinary Green's function G, we need to start from the following homogeneous boundary conditions. The homogeneous equation with the operator L becomes

$$Lu(x) \equiv -[\nabla^2 + q(x)]u(x) = 0 \quad (x \in V) \tag{5.69}$$

The Green's function G satisfies

$$-[\nabla^2 + q(x)]G(x, x') = \delta(x - x') \tag{5.70}$$

It is known from Kirchhoff formula that

$$u(x) = \iint\limits_{S} \left[G(x, x') \frac{\partial u(x)}{\partial n} - u(x') \frac{\partial G(x, x')}{\partial n} \right] dS \tag{5.71}$$

The solution to the direct problem of the wave equation should satisfy Eqns (5.69), (5.71) and (5.66) simultaneously. For Neumann boundary condition

$$\frac{\partial G(x, x')}{\partial n} = 0 \quad (x \in S),$$

it should hold that

$$G_S(x, x') = G(x, x'). \tag{5.72}$$

For Dirichlet boundary condition

$$Bu \equiv u(x) = b(x) \quad (x \in S),$$

we should have

$$G_S(x, x') = -\frac{\partial G(x, x')}{\partial n} \tag{5.73}$$

From the results, we can find the transformation relationship between the second type Green's function G_s and the Green's function G according to the different types of the boundary conditions.

5.7. SUMMARY

By using finite element method or finite difference method, we can compute the synthetic seismogram accurately. But in order to understand the insight of the wave propagations, we need to find the analytical solution with specified initial and boundary values in some cases. For example, the reflected wave field by a surface of finite size is composed of the reflected and diffracted waves. On the stacked sections, the reflected and diffracted waves exist at the same time. Reflected wave comes from the reflectors that correspond to the soft boundary conditions. Diffracted wave comes from diffractors that correspond to the termination of the soft boundary conditions, which is also called the singularity in mathematics. Seismic reflection interpretation involves two matters: the accurate positioning and quantitative estimation of the properties of both the reflectors and the diffractors. The first two sections of this chapter show that the analytical solution can be obtained for some initial and boundary value problems of the wave equation with constant coefficients. If the coefficients of the wave equation are as usual not constant, but change in a slow, continuous, and smooth manner, we can find the approximate solution of the wave equation by applying the Green's function method. In Sections 5.4—5.6, we have discussed the principle of the Green's functions as well their expression details. However, for the initial and boundary value problems with variable coefficients in the wave equation, the solution cannot be obtained by employing simple extension from the problems of a constant coefficient wave equation. We must study some more generalized operators with their theory being more complex. In the next chapter, we will learn the Fourier integral operators and the pseudo-differential operators.

6

Decomposition and Continuation of Seismic Wave Field

In the previous chapter, we have discussed the solutions of the wave equation with boundary and initial value conditions. In this chapter, we will discuss the decomposition and extrapolation of seismic wave field. We should discuss the inverse problem of the wave equation in the following chapter. By discussing the decomposition and extrapolation of seismic wave field, we introduce a mathematic ideal called the expansion of wave differential operators. Although it is not easy to explain clearly the expansion of wave differential operators for engineers, the expansion ideal has wide applications in engineering. In matrix operation, we have discussed the singular value decomposition method for solving difficult linear equations, which is a good example of the numerical expansion of differential operators. Insight of the expansion of wave operators in inhomogeneous media is very useful for understanding the nature of wave propagation.

The characteristic of reflection wave field is that the ray's angle between reflector and the source and the ray's angle between the reflector and the receiver are never obtuse, thus the reflection can be recorded on the surface. For the shallow borders of seabed or soil layer, this angle can be obtuse yet, but the shallow borders are usually not the targets for seismic reflection exploration. As mentioned in Chapter 4, the reflections from the shallow layers are often removed in seismic data processing. Seismic reflection methods are only applied to locate echoes from the deeper underground layers, which generate the upgoing waves. In reflection data processing, we eliminate the downgoing wave directly generated by seismic sources, because it will disturb imaging reflectors produced by underground interfaces or thin layers. Because the ray's angle between the reflector and the source and the ray's angle between the reflector and the receiver are both acute, almost no wave signal traveling close to the horizontal can be recorded in the reflection method. Therefore, the decomposition of the upgoing wave and the downgoing wave is an important matter in the reflection seismology.

Because the target of the seismic reflection is located deep underground, there is the problem of extrapolation of the seismic field observed at the surface to a greater depth, so that the reflection wave field becomes closer to subsurface targets. Mathematically, this problem belongs to downward wave continuation, which is different from the problem of seismic wave migration. The goal of the wave migration is to relocate the wave field exactly to the moment when the excitation just occurred in the subsurface by employing wave field backpropagation, thus compensating for the distortion of reflector's location caused by the wave propagation. The depth change in the seismic wave migration is not a constant, while the depth change used in the seismic continuation procedure is a constant. Although the downward seismic continuation belongs to the direct problems of wave equation with initial and boundary value conditions, it is similar to the inverse problems in the sense that they are both ill-posed problems in mathematics. Chinese mathematicians Luan Wengui et al. (1989) achieved remarkable results on this subject. Before discussing the inverse problems of the wave equation in the next chapter, we will first discuss the downward continuation of the seismic wave field in this chapter.

6.1. THE EQUATIONS OF ACOUSTIC UPGOING AND DOWNGOING WAVES

Let us first study the upgoing and downgoing waves. Quantitatively, they should be fluctuation signals that can be defined accurately by equations. However, be careful as there are two different definitions for the upgoing and downgoing waves. For the first, the upgoing and downgoing waves are defined by order reduction of differential equations that can be applied to continuous media with varying velocities. The 1D homogeneous acoustic wave equation with variant coefficient can be written as

$$\frac{1}{c^2(z)}\frac{\partial^2 u}{\partial t^2} - \frac{\partial^2 u(z,t)}{\partial z^2} = 0 \tag{6.1}$$

Letting the downward direction of Z-axis to be positive, the definition of the acoustic downgoing wave after the order reduction is (Luan Wengui, 1989)

$$D(z,t) = \frac{1}{2}\left[\frac{1}{c(z)}\frac{\partial u}{\partial t} - \frac{\partial u}{\partial z}\right] \tag{6.2a}$$

and the definition of the acoustic upgoing wave after the order reduction is:

$$U(z,t) = \frac{1}{2}\left[\frac{1}{c(z)}\frac{\partial u}{\partial t} + \frac{\partial u}{\partial z}\right] \tag{6.2b}$$

We will see that if the upgoing and downgoing acoustic waves defined in Eqns (6.2a) and (6.2b) satisfy the pseudo-differential equations:

$$\frac{1}{c(z)}\frac{\partial D}{\partial t} + \frac{\partial D}{\partial z} = -\frac{c'(z)}{2c(z)}(D+U) \tag{6.3a}$$

and

$$\frac{1}{c(z)}\frac{\partial U}{\partial t} - \frac{\partial U}{\partial z} = -\frac{c'(z)}{2c(z)}(D-U) \tag{6.3b}$$

where $c'(z)$ is the vertical varying rate of the velocity, then Eqns (6.3a) and (6.3b) satisfy the wave Eqn (6.1) and the upgoing and downgoing waves are equivalent to the wave field u.

Actually, from Eqn (6.2), we have

$$D + U = \frac{1}{c(z)} \frac{\partial u}{\partial t}$$

$$D - U = -\frac{\partial u}{\partial z} \tag{6.4}$$

Substituting Eqn (6.4) into Eqns (6.3a) and (6.3b) and adding them, we have

$$\frac{1}{c} \frac{\partial D}{\partial t} + \frac{\partial D}{\partial z} + \frac{1}{c} \frac{\partial U}{\partial t} - \frac{\partial U}{\partial z} = 0$$

Substituting Eqn (6.4) into the above equation, we can obtain the wave Eqn (6.1). Thus it is proved that there is equivalence between the equations of upgoing and downgoing waves in Eqns (6.3a) and (6.3b) and the total wave field u in Eqn (6.1).

It is important to notice that the upgoing and downgoing waves discussed above are called "order-reduced upgoing and downgoing waves", which is completely different from upward and downward waves commonly used by engineers for analysis of reflection or vertical seismic profile (VSP) diagrams. Equations (6.3) and (6.4) reveal the law of movement of acoustic waves by using terms of the upgoing and downgoing waves, constraint to the cases of plane wave incidence. This law suits both reflection seismic and VSP records. In seismic interpretation of VSP and cross-hole seismic seismograms, engineers easily distinguish between the common upward and downward waves by looking at the slope of the wave chains. The concepts of the upward and downward waves come from comparison of locations of receivers and shots. In accurate analysis, we show "order-reduced upgoing and downgoing waves" instead, which are only relative to the source of incident plane wave of going downward, but not related to the observation arrangement.

Now let us discuss two-dimensional "order-reduced upgoing and downgoing waves". During acoustic wave propagation, wave field $u(x,z,t)$ satisfies the homogeneous wave equation as

$$\frac{1}{c^2} \frac{\partial^2 u}{\partial t^2} - \frac{\partial^2 u}{\partial x^2} - \frac{\partial^2 u}{\partial z^2} = 0 \tag{6.5}$$

where c is the acoustic wave velocity and is assumed to be constant. After expressing the above equation in the frequency–wave number $(f–k_x)$ domain, it becomes

$$\left(\frac{\partial^2}{\partial z^2} + k_z^2 \right) u(k_x, z, \omega) = \left(\frac{\partial}{\partial z} + ik_+ \right) \left(\frac{\partial}{\partial z} - ik_- \right) u(k_x, z, \omega) = 0 \tag{6.6}$$

where

$$k_x^2 + k_z^2 = \frac{\omega^2}{c^2}$$

$$k_{\pm} = \pm k_z = \pm \frac{\omega}{c}\sqrt{1 - \frac{k_x^2 c^2}{\omega^2}}$$

(6.7)

where "+" corresponds to the upgoing wave vector $+\mathbf{k}$, and "−" corresponds to the downgoing wave vector $-\mathbf{k}$. However, Eqns (6.6) and (6.7) are only valid for Eqn (6.5) with constant velocity medium. As the purpose of Earth exploration is to detect the inhomogeneous properties of the medium, the wave velocity must be assumed to be variant and the "order-reduced upgoing and downgoing waves" should be studied.

Referring to Eqns (6.6) and (6.7), the order-reduced downgoing wave can be defined as:

$$\frac{\partial D(k_x, z, \omega)}{\partial t} = -\frac{1}{2}\left[\frac{\partial u(k_x, z, \omega)}{\partial z} + ik_- u(k_x, z, \omega)\right]$$

(6.8a)

While the order-reduced upgoing wave can be defined as:

$$\frac{\partial U(k_x, z, \omega)}{\partial t} = \frac{1}{2}\left[\frac{\partial u(k_x, z, \omega)}{\partial z} + ik_+ u(k_x, z, \omega)\right]$$

(6.8b)

If the order-reduced upgoing and downgoing waves satisfy the pseudo-differential Eqns (6.8a) and (6.8b), we have after using Fourier transform for constant velocity

$$\frac{\partial D(x, z, t)}{\partial t} = \frac{1}{2\pi}\iint D_t(k_x, z, \omega)e^{i\omega t + ik_x x}d\omega\, dk_x$$

(6.9a)

and

$$\frac{\partial U(x, z, t)}{\partial t} = \frac{1}{2\pi}\iint U_t(k_x, z, \omega)e^{i\omega t + ik_x x}d\omega\, dk_x$$

(6.9b)

Then by using Eqn (6.6), we get the equations for the order-reduced upgoing and downgoing waves

$$\left(\frac{\partial}{\partial z} + ik_+\right)D_t(k_x, z, \omega) = 0$$

(6.10a)

and

$$\left(\frac{\partial}{\partial z} + ik_-\right)U_t(k_x, z, \omega) = 0$$

(6.10b)

One can see that the equations satisfied by the order-reduced upgoing and downgoing waves Eqn (6.10) are equivalent to the wave Eqn (6.6).

In practice, the wave velocity c is not a constant and the equations of using Fourier transform Eqn (6.9) should not be used for the accurate definition of the order-reduced upgoing and downgoing waves. In these cases, we must deal with the theory of pseudo-differential operators, which is a natural generalization of the Fourier transform. Assuming $c(x,y)$ to be a smooth function, an approximation of the solution for wave of high frequencies can be derived by using the theory of pseudo-differential operators. The solution is in harmonic function form, and its amplitude function and phase function can be derived as the eikonal equation and transport equations introduced in Section 5.5. The problem of decomposition of the upgoing and downgoing waves in three-dimensional (3D) space will be discussed later in Section 6.6.

Equations (6.1)–(6.4) show that the one-dimensional wave field can be separated into upgoing and downgoing waves but that the wave field is not simply the sum of the two. Thus, it is necessary to distinguish them from the upward and downward waves in VSP records. From Eqn (6.4), it can be seen that the sum of the order-reduced upgoing and downgoing waves equals the time differential of the wave field divided by the wave velocity; while the difference between the order-reduced upgoing and downgoing waves equals the partial derivative of the wave field with respect to the z-axis. Equation (6.3) satisfied by the order-reduced upgoing and downgoing waves are pseudo-differential equations of the first order, which are one order less than that of the common wave equations. The order-reduced upgoing and downgoing waves are coupled to each other in Eqn (6.3). In order to derive the upgoing and downgoing waves, we need to know the partial derivative of the wave field with respect to z, except the time derivative and wave velocity. Usually, the partial derivative of the wave field with respect to z cannot be observed directly. The continuation of the wave field is a good method for computing the partial derivative.

6.2. KIRCHHOFF MIGRATION OF THE PRESTACK SEISMIC DATA

We have discussed seismic wave migration for zero-offset reflection gathers. In this section, we will discuss the application of the Kirchhoff integral formula to the continuation of the acoustic wave field, providing a theoretical basis to the prestack migration of wave field. As discussed in Chapter 5, given the wave field u at an enclosed border and the Green's function G of the wave equation in half space with a point source, the

Kirchhoff integral formula for calculating the wave field in the medium within the border is

$$u(x,t) = \frac{1}{4\pi} \int dt' \iint_S \left\{ g(x,x',t') \frac{\partial u(x,x')}{\partial n} - u(x,t) \frac{\partial g(x,x',t')}{\partial n} \right\} dS_{x'} \quad (5.8b)$$

The application of the formula to prestack wave migration in reflection seismology is introduced as follows.

The domain studied in the prestack wave migration is the lower half space, i.e. the underground of the Earth. The surface of the earth at $z = 0$ acts as a mirror in reflection wave propagation, producing a correction term to the integral in Eqn (5.8b). The correction term represents a new seismic phase from the reflective mirror with an imaginary source located above the ground surface. Correspondingly, we should add a term to the Green's function as (Frittman, 1965)

$$g(r;t,r',t') = \frac{\delta\left(t - t' - \frac{r}{c}\right)}{r} - \frac{\delta\left(t - t' - \frac{r'}{c}\right)}{r'} \quad (6.11)$$

where r is the distance from a point (x,y,z) in the wave field to the real point source (x_0,y_0,z_0) and r' is the distance from the point (x,y,z) to the imaginary point source above the ground:

$$r = \sqrt{(z - z_0)^2 + (x - x_0)^2 + (y - y_0)^2}$$

$$r' = \sqrt{(z + z_0)^2 + (x - x_0)^2 + (y - y_0)^2}$$

When the real source is on the ground ($z_0 = 0$) and superimposed with the imaginary source, noticing that the normal direction of the ground is the vertical axis, Eqn (5.8b) can be written as:

$$u(r,t) = \frac{1}{2\pi} \int \left\{ \iint_S u(r',t') \frac{\partial}{\partial z_0} \left[\frac{\delta\left(t' - t - \frac{r}{c}\right)}{r} \right] dS \right\} dt' \quad (6.12)$$

where $u(r',t)$ is the wave field on the ground. It is also an integral formula of Kirchhoff with elimination of the partial derivative term of u in Eqn (5.8b), which makes wave field extrapolation possible. If we have smooth ground surface, Eqn (6.12) can also be used for approximate computation.

Because Eqn (6.12) takes the form of convolution integral, we can use it for the extrapolation of wave field from plane $z = z_0$ to plane $z > 0$; this method is called the poststack phase-shift migration. Rewrite the convolution integral form of Eqn (6.12) as

$$u(x, y, z, t) = \frac{1}{2\pi} u(x, y, z_0, t) * \frac{\partial}{\partial z_0} \left\{ \frac{\delta \left(t \pm \frac{r}{c} \right)}{r} \right\} \tag{6.13}$$

where "*" denotes the convolution operator. Letting $\Delta z = z - z_0$, we get

$$r = \sqrt{\Delta z^2 + (x - x_0)^2 + \left(y - y_0 \right)^2}$$

Assuming the wave velocity to be constant, using Fourier transform with respect to x, y, t, in the frequency−wave number domain, Eqn (6.13) can be written as

$$u\left(k_x, k_y, z, \omega \right) = u\left(k_x, k_y, z_0, \omega \right) H\left(k_x, k_y, \Delta z, \omega \right) \tag{6.14}$$

where the linear operator is

$$H\left(k_x, k_y, \Delta z, \omega \right) = e^{\pm i |\Delta z|} \sqrt{\frac{\omega^2}{c^2} - k_x^2 - k_y^2} \tag{6.15}$$

These equations give the poststack phase-shift migration formula (Berkout, 1985). In Eqn (6.15), sign "+" represents downward continuation of the wave field, while the sign "−" represents upward continuation.

The poststack migration formula can only be used theoretically to the continuation of plane wave field. The condition for prestack downward continuation is different, as both the point shot and point receivers need to be shifted downward; sometimes the shifting of the shot points and receiving points needs to be considered at the same time. At the surface, $z = z_0 = 0$ and the shots $(x_{s0}, y_{s0}, 0)$ and the receiving points $(x_{r0}, y_{r0}, 0)$ are marked with the subscript "0", while on the continuation plane, $dz > 0$ and the shot point (x_{s1}, y_{s1}, dz) and the receiving point (x_{r1}, y_{r1}, dz) are marked with the subscript "1". Let us consider records of the common-shot gather first. We can try to move the receivers downward and keep the shot stay. By using Eqn (6.13), we have

$$u\left(r_1, r_0', t \right) = -\frac{1}{c} \iint\limits_{S_0} \frac{\partial R}{\partial z_0} \frac{1}{R} \frac{\partial}{\partial t} u\left(r_0, r_0', t + \frac{R}{c} \right) dS_0 \tag{6.16}$$

where S_0 is the observation surface, r_0' is the distance from the surface shot to the origin, and r_0 is the distance from the receiving points $(x_{r0}, y_{r0}, 0)$ on the surface to the origin. Denoting r_1 as the distance from the downward moving point (x_{r1}, y_{r1}, dz) in the continuation plane to the origin, we get

$$r_1 = \sqrt{\Delta z^2 + x_{r1}^2 + y_{r1}^2} \tag{6.17a}$$

$$R = |r_1 - r_0| \tag{6.17b}$$

What we have to do next is to move the shot points downward from records of the common-receiver gathers. Let r'_1 be the distance between the downward moving shot points and the origin:

$$r'_1 = \sqrt{\Delta z^2 + x_{s1}^2 + y_{s1}^2} \tag{6.18a}$$

$$R' = |r'_1 - r'_0| \tag{6.18b}$$

Once more, the computation of the integration of the wave field along the shot points gives

$$u(r_1, r'_1, t) = -\frac{1}{c} \iint_{S1} \frac{1}{R'} \frac{\partial R'}{\partial z_0} \frac{\partial}{\partial t} u\left(r_1, r'_0, t + \frac{R'}{c}\right) dS_1 \tag{6.19}$$

where S_1 is the continuation plane $z = \Delta z$. Substituting Eqn (6.16) into Eqn (6.19), we have the downward wave continuation formula for prestack seismic migration as

$$u(r_1, r'_1, t) = -\frac{1}{c^2} \iint_{S_0} dS_0 \iint_{S1} \frac{1}{RR'} \frac{\partial R}{\partial z_0} \frac{\partial R'}{\partial z_0} \frac{\partial^2}{\partial t^2} u\left(r_0, r'_0, t + \frac{R + R'}{c}\right) dS_1 \tag{6.20}$$

The purpose of seismic wave migration is to relocate the wave field to the moment when the excitation just occurs from reflectors, in order to eliminate distortion of the reflector's image coursed during the wave propagation. The ideal method should be the backpropagation of wave field, to which the downward continuation belongs. As stated in Section 4.7, we can find the migrated wave field by letting $r_1 = r'_1$ and $t = 0$ in Eqn (6.20). The depth of the backpropagation in prestack wave migration is constrained by $r_1 = r'_1$, where $t = 0$ means that the wave field has been relocated to the right position corresponding to the excitation moment.

6.3. DOWNWARD CONTINUATION OF THE REFLECTIVE SEISMIC WAVE FIELD IN HOMOGENOUS MEDIA

Let us first discuss the simplest case of homogenous medium, when there are analytic solutions to the downward continuation problem of the reflected wave field. Given the wave field and its space derivative on an enclosed border, the wave propagation within the boundary can be computed by using the Kirchhoff formula. Downward continuation is defined as finding the wave field $u(x,y,z < 0,t)$ given the wave field $u(x,y,0,t)$ at the surface $z = 0$. In homogenous medium, the downward continuation problem becomes a boundary value problem defined as follows:

$$\frac{\partial^2 u}{\partial z^2} + \left(\frac{\partial^2}{\partial x^2} + \frac{\partial^2}{\partial y^2} - \frac{1}{c^2}\frac{\partial^2}{\partial t^2} \right) u = 0, -H \le z < 0 \tag{6.21}$$

$$u\big|_{z=0} = f(x, y, t), \frac{\partial u}{\partial t}\bigg|_{z=0} = g(x, y, t)$$

where H is the maximum depth of the continuation and c is constant velocity. Equation (6.21) is a mixed Cauchy boundary value problem. We will prove in the following discussion that the vertical derivative g can be derived from the given wave field f on the boundary.

Applying Fourier transform on both sides of Eqn (6.21) and letting wave number $k_x = \xi$, $k_y = \eta$, the solution of Eqn (6.21) can be written in the wave number–frequency domain as (Luan Wengui, 1989; Yang Wencai, 1989)

$$u(z, x, y, t) = \frac{-1}{(2\pi)^{3/2}} \int_0^\infty d\omega \int\int_{\xi^2+\eta^2 \le \omega^2/c^2} u(z, \xi, \eta, \omega)e^{i(x\xi+y\eta+t\omega)} d\xi\, d\eta \tag{6.22}$$

where the space integral domain satisfies:

$$\xi^2 + \eta^2 < \frac{\omega^2}{c^2} \tag{6.23a}$$

For downward continuation, we have $-H < z < 0$, and for upward extrapolation, we have $z > 0$. Denoting parameters in the wave number–frequency domain by adding "$-$" on their top and applying Fourier transform to Eqn (6.21), we have for upward continuation,

$$\bar{g}(\xi, \eta, \omega) = -\sqrt{\xi^2 + \eta^2 - \frac{\omega^2}{c^2}}\bar{f}(\xi, \eta, \omega) \tag{6.23b}$$

The first boundary condition in Eqn (6.21) can be rewritten in the wave number–frequency domain as

$$\bar{m}(\xi, \eta, \omega) = \begin{cases} \bar{f}(\xi, \eta, \omega), & \xi^2 + \eta^2 \ge \frac{\omega^2}{c^2} \\ 0, & \text{others} \end{cases} \tag{6.24}$$

Substituting it to the second boundary condition Eqn (6.23b) and applying the reverse Fourier transform, we have

$$n(z, x, y, t) = \frac{-1}{(2\pi)^{3/2}} \int_0^\infty d\omega \int\int_{\xi^2+\eta^2 \le \omega^2/c^2} \sqrt{\xi^2 + \eta^2 - \frac{\omega^2}{c^2}}\bar{m}(\xi, \eta, \omega)e^{i(x\xi+y\eta+t\omega)} d\xi\, d\eta \tag{6.25}$$

From the results, we prove in the constant velocity medium that the two boundary conditions in Cauchy problem Eqn (6.21) can be combined and represented by the first boundary condition.

If one would like to perform upward continuation by using the sub-surface wave field $u(x,y,z,t)$ to find the surface wave field $u(x,y,0,t) = f(x,y,t)$, we have

$$\iiint K_z(x - x', y - y', t - t')u(z, x, y, t)dx'\, dy'\, dt' = f(x,y,t), \quad -H \leq z < 0$$

(6.26)

where the kernel function K_z is similar to Green's function that can be derived from Eqn (6.15):

$$K_z(z, x - x', y - y', t - t')$$
$$= \frac{-1}{8\pi^3} \int\limits_0^\infty d\omega \iint\limits_{\xi^2+\eta^2 \leq \omega^2/c^2} e^{-z\sqrt{\xi^2+\eta^2-\frac{\omega^2}{c^2}}+i(\xi(x-x')+\eta(y-y')+\omega(t-t'))}d\xi\, d\eta \qquad (6.27)$$

Now we consider downward continuation as the reversal to the upward continuation. We want wave field $u(x,y,z,t)$ from the surface wave field $u(x,y,0,t) = f(x,y,t)$. The operator for downward continuation operator must be the inverse operator of the upward extrapolation operator. In order to get it, we just have to reverse the sign of the exponential term in the kernel function of Eqn (6.27)

$$K_d(z, x - x', y - y', t - t')$$
$$= \frac{-1}{8\pi^3} \int\limits_0^\infty d\omega \iint\limits_{\xi^2+\eta^2 \leq \omega^2/c^2} e^{z\sqrt{\xi^2+\eta^2-\frac{\omega^2}{c^2}}+i(\xi(x-x')+\eta(y-y')+\omega(t-t'))}d\xi\, d\eta \qquad (6.28)$$

where $-H \leq z < 0$. We have completed the derivation of both upward and downward continuation for the case of constant velocity. The computation of downward continuation is finding the kernel function by using Eqn (6.28) first and then substituting it into Eqn (6.26), which belongs to the first-kind Fredholm integral equation. Finally, we solve this equation and find a numerical solution for it.

Looking back again at the conditions in integral domain Eqn (6.23a), it can be seen that the integral condition Eqn (6.23) in wave number—frequency domain is important and must be paid attention to during seismic inversion and reflection data processing. The domain boundary defined by the inequality is:

$$\xi^2 + \eta^2 = \frac{\omega^2}{c^2} \tag{6.29}$$

Equation (6.29) defines a conical surface in wave number—frequency domain. Let $\varepsilon > 0$ be a small number; the conical surface is the border of three regions in the wave number—frequency domain as follows:

1. The horizontal sweeping region that is located close to the conical surface, where

$$\xi^2 + \eta^2 + \frac{\omega^2}{c^2} \leq \varepsilon$$

2. The reflection diminishing region that is located outside the conical surface, where

$$\xi^2 + \eta^2 + \frac{\omega^2}{c^2} > \varepsilon$$

3. The vertical reflecting region is inside the conical surface, satisfying condition Eqn (6.23) and corresponding to the integral domain. Hence in reflection seismology, the angle between the source ray from a reflector and the receiver ray from the reflector must be acute, while the conical surface is symmetrical to the z-axis. So, any computation procedure in reflection seismic processing must satisfy the condition of the vertical reflecting region.

When downward continuation is performed as a direct problem to compute $u(x,y,z,t)$, we have to substitute Eqn (6.28) into the integral:

$$\iiint K_d(z, x - x', y - y', t - t')f(x', y', t')dx' \, dy' \, dt' = u(z, x, y, t), -H \leq z < 0 \tag{6.30}$$

where K_d is the inverse operator in Eqn (6.28) and f is the first boundary condition of the wave field in Eqn (6.24). Because the integral kernel shown in Eqn (6.28) is an exponential function, which increases rapidly with the increase of the extrapolation depth H, the calculation using Eqn (6.26) can be unstable. Downward continuation problem belongs to the so-called ill-posed problems in mathematics. The so-called ill-posed problems are hard to be solved exactly because their definition is not accurate or perfect (Tiknonov and Arsenin, 1977; Garnir, 1980; Marozov, 1984). In order to ensure the stability of the computation of the downward continuation procedure, mathematicians introduce the following functional equations for approximate solutions (Aarts and Korst, 1989) as mentioned in Section 3.1:

$$J_1(u) \equiv \iiint \|u(x,y,-H,\omega)\|^2 dx\, dy\, d\omega = \min \tag{6.31}$$

$$J_2(u) \equiv \iiint \left\| \widehat{f} - f \right\|^2 dx\, dy\, d\omega = \min \tag{6.32}$$

where $\widehat{f}$ is the synthetic wave field at the surface computed by using direct method from Eqn (6.26) and K_z. K_z is the integral kernel of upward continuation shown in Eqn (6.27), f is the boundary condition Eqn (6.24) that corresponds to wave field observed on the surface $z = 0$. In fact, the functionals (6.31) and (6.32) lead to a contradiction if they require both Eqns (6.31) and (6.32) to be worked; that is why the problem is called "ill-posed". A better way to improve the procedure is requiring

$$J_2 + \alpha J_1 = \min \tag{6.33}$$

where "min" means minimum value of the functional and α is called the regularization factor or compromise coefficient. The regularization method takes the second-order norm, i.e. $\| * \|$, of the functional in (6.31) and (6.32), while the Backus and Gilbert method (Backus and Gilbert, 1967, 1970) takes the first-order norm, i.e. $| * |$, of the functional in (6.31) and (6.32).

After regularization of the functional (6.33), the formula for wave field downward continuation becomes (Luan Wengui, 1989):

$$u(x,y,z,t) = \frac{1}{\pi^3} \int\!\!\!\int\!\!\!\int_0^\infty \varphi(z,\xi,\eta,\omega) d\xi$$

$$d\eta\, d\omega \int\!\!\!\int\!\!\!\int_{-\infty}^\infty f(x',y',t')\cos(\xi(x-x'))\cos(\eta(y-y'))\cos(\omega(t-t'))dx'\, dy'\, dt'$$

$$\tag{6.34}$$

where $-H \le z < 0$ and

$$\varphi(z,\xi,\eta,\omega) = \begin{cases} \dfrac{e^{-z\sqrt{\xi^2+\eta^2-\frac{\omega^2}{c^2}}}}{1+\alpha e^{-z\sqrt{\xi^2+\eta^2-\frac{\omega^2}{c^2}}}}, & \xi^2 + \eta^2 \ge \dfrac{\omega^2}{c^2} \\[2em] e^{-z\sqrt{\xi^2+\eta^2-\frac{\omega^2}{c^2}}}, & \xi^2 + \eta^2 < \dfrac{\omega^2}{c^2} \end{cases} \tag{6.35}$$

taking depth z negative means downward continuation. These formulas give the solution to downward continuation, which is a Cauchy mixed boundary value problem.

6.4. DOWNWARD CONTINUATION OF SEISMIC WAVE FIELD IN VERTICALLY INHOMOGENEOUS MEDIA

Let us first discuss a case which is a little more complex than the homogeneous medium mentioned above. The analytic solution for the downward continuation problem of reflected wave field can be found under certain conditions when wave velocity of the medium changes only with depth. In the vertically inhomogeneous and transversely homogeneous medium, homogeneous wave equation satisfies the acoustic equation with a pseudo-differential operator:

$$\nabla^2 u - \frac{1}{c(z)^2} \frac{\partial^2 u}{\partial t^2} = 0 \tag{6.36}$$

where $u = u(x,y,z,t)$. The seismic downward continuation problem becomes a mixed Cauchy boundary value problem, which is to derive $u(x,y,z,t)$ given

$$u(x, y, 0, y) = f(x, y, t); \quad \partial u(x, y, 0, t)/\partial t = g(x, y, t) \tag{6.37}$$

where H denotes the maximum continuation depth, $-H \le z < 0$, and $z = 0$ represents the ground surface. After assuming that the given wave field $u(x,y,0,t)$ and its derivative on the boundary are continuous, and that the wave velocity on the surface is $c(z = 0) = c(0)$, similar to the derivation of Eqn (6.25), the derivative g can be derived from f:

$$g(x, y, t) = \frac{-1}{(2\pi)^{3/2}} \int\limits_{-\infty}^{\infty} d\omega \int\limits_{-\infty}^{\infty} \int\limits_{-\infty}^{\infty} \sqrt{\xi^2 + \eta^2 - \omega^2/c^2(0)} f(\xi, \eta, \omega) e^{i(x\xi + y\eta + t\omega)} d\xi \, d\eta.$$

$$\tag{6.38}$$

In the wave number−frequency domain, we have:

$$\widehat{g}(\xi, \eta, \omega) = -\sqrt{\xi^2 + \eta^2 - \omega^2/c^2(0)} \widehat{f}(\xi, \eta, \omega), \tag{6.39}$$

If velocity function $c(z)$ is not set appropriately, there would be no analytic solution for the problem defined in Eqns (6.36)−(6.38). However, under some specific conditions, Luan (1989) has derived corresponding explicit solutions, although the conditions of vertical velocity change used are not completely suitable in seismic exploration. We will next discuss the medium of velocity varying linearly with depth, as in this case, the downward continuation problem is more suitable for seismic exploration.

In Section 5.3, we have discussed the Green's function in medium with linear velocity-depth functions, which can be used to solve the problem of downward continuation here. Suppose that in a Cartesian coordinate system the acoustic velocity that changes in the vertical direction is denoted by

$$c(z) = a + bz, \ b > 0$$

where $a = c(0)$ denotes surface velocity and b denotes vertical velocity gradient. Because wave velocity is not constant, coordinate transformation is necessary. Introducing the new vector τ with three components (Huang Liang-Jiu and Yang Wencai, 1991), we get

$$\tau_1 = \frac{x}{c(z)}$$

$$\tau_2 = \frac{y}{c(z)}$$

$$\tau_3 = \int_0^z \frac{1}{c(z')} dz'$$

It is called pseudo-time coordinate vector. Substituting it into Eqn (5.26), we have

$$\tau_3 = \frac{1}{b} \ln\left(\frac{a + bz}{a}\right) = \frac{1}{b} \ln\left(\frac{c(z)}{a}\right)$$

$$c(\tau_3) = a e^{b\tau_3}$$

In the ground surface $z = 0$, we have $\tau_3 = 0$ and $c(\tau_3) = c(0) = a$. For the subsurface extrapolation plane $z = h$, we have

$$\tau_3 = \frac{1}{b} \ln\left(\frac{a + bh}{a}\right) \tag{6.40}$$

In Cartesian coordinates, a point is denoted by $x = (x,y,z)$; the harmonic wave equation in the frequency domain with excitation of a point source is:

$$\left(\nabla^2 + \frac{\omega^2}{c^2(z)}\right) u(x, x', \omega) = -\delta(x - x')$$

The left side of the equation is not an ordinary partial differential operator, but a pseudo-differential operator. The downward continuation problem is to derive the wave field $u(x,t)$ on subsurface $x = (x,y,h)$ given the wave field $u(x',t)$ on the surface $x' = (x',y',0)$.

As derived in Section 5.3, the acoustic wave equation in the pseudo-time coordinates becomes

$$\left(\nabla^2 - b\frac{\partial}{\partial \tau_3} + \omega^2 \right) u(\tau, \tau', \omega) = -\frac{1}{a}\delta(\tau - \tau')$$

The downward continuation problem derives the subsurface wave field $u_h(\tau,t)$ given the wave field $u_0(\tau',t')$ on the ground $\tau'_3 = 0$. The Green's function in the medium with velocity of varying linearly along z is

$$\left(\nabla^2 - b\frac{\partial}{\partial \tau_3} + \omega^2 \right) G(\tau, \tau', \omega) = -\delta(\tau - \tau')$$

Comparing the above two equations, we have

$$u(\tau, \tau', \omega) = \frac{1}{a} G(\tau, \tau', \omega)$$

The wave field u is proportional to its Green's function G. In contrast to the homogenous medium, the coefficient becomes $(1/a)$.

As derived in Section 5.3, the Green's function corresponding to Eqn (5.36b) is:

$$G(\tau, \tau', \omega) = \frac{e^{\frac{b\tau_3}{2}}}{4\pi|\tau - \tau'|} e^{\mp\sqrt{\omega^2 - \left(\frac{b}{2}\right)^2}|\tau - \tau'|}$$

When the wave velocity gradient is not large, assume

$$\frac{b}{2} \ll \omega \quad \text{then} \quad \sqrt{\omega^2 - \left(\frac{b}{2}\right)^2} \approx |\omega|,$$

The high-frequency approximation of the Green's function becomes

$$G(\tau, \tau', \omega) = \frac{e^{\frac{b\tau_3}{2}}}{4\pi|\tau - \tau'|} e^{\mp\omega|\tau - \tau'|}$$

Based on the Green's function solution (5.40) and (6.30), the solution to the downward continuation problem of the wave field can be written as

$$u_h(\tau, \omega) = \frac{1}{4\pi} \int\int_{-\infty}^{\infty} \frac{e^{\frac{b\tau_3}{2}}}{|\tau - \tau'|} e^{i\sqrt{\omega^2 - \left(\frac{b}{2}\right)^2}|\tau - \tau'|} u_0(\tau', \omega) d\tau'_1 \, d\tau'_2 \qquad (6.41)$$

where $\tau = (\tau_1, \tau_2, \tau_3)$ and τ_3 is defined in Eqn (6.40). For the ground surface $\tau_3' = 0$, we have

$$|\tau - \tau'| = \sqrt{(\tau_1 - \tau_1')^{2} + (\tau_2 - \tau_2')^2 + h^2} \tag{6.42}$$

Based on the inverse operator of the Green's function (5.40), the solution to the upward continuation problem of the wave field can be written as:

$$u_{-h}(\tau, \omega) = \frac{1}{4\pi} \int\limits_{-\infty}^{\infty}\!\!\int \frac{e^{-\frac{b\tau_3}{2}}}{|\tau - \tau'|} e^{-i\sqrt{\omega^2 - \left(\frac{b}{2}\right)^2}\,|\tau - \tau'|} u_0(\tau', \omega)\, d\tau_1'\, d\tau_2' \tag{6.43}$$

where τ_3 is defined in Eqn (6.40) and the extrapolation depth h must take negative value. Equation (6.43) gives the solution for the downward continuation problem, having limited applications due to the assumption that the underground velocity must be linear functions with depth as defined in Eqn (5.31).

In short, the computation of the wave field downward continuation in medium with velocity varying linearly with depth contains three steps. First, convert the Cartesian coordinates into the pseudo-time coordinates by using Eqns (5.33)–(5.35). Second, compute the downward continuation wave field by using Eqn (6.43) in the pseudo-time coordinate system. Finally, convert the wave field function in the pseudo-time coordinates back to Cartesian coordinates by performing interpolation to get the wave field $u(x,y,z < 0,t)$.

6.5. THE PSEUDO-DIFFERENTIAL OPERATOR AND FOURIER INTEGRAL OPERATOR

Here after, we will discuss the downward continuation problem of the reflected wave field in inhomogeneous medium, where the wave velocity $c(x,y,z)$ in the wave equation is a function of 3D space variants. New mathematical tools need to be developed for solving the boundary value problem of wave equation with variant coefficients. In this section, we will mainly introduce a new mathematics tool, namely, the expansion of Fourier integral operator. By using the expansion, we will discuss the downward continuation problem of reflected seismic wave field in inhomogeneous medium in the next section.

6.5.1. Analysis of the Boundary Value Problem of Wave Equation with Variant Coefficients

In the above discussion, we have been employing the classical mathematics tools to solve the constant coefficient wave equation and avoided answering directly the question what are the properties of the solution of the wave equation with variant coefficients? Are there proper mathematical tool for the boundary value problems of the wave equation with variant coefficients? Obviously, these problems are very important in theoretical study of geophysical inversion and Earth structure imaging. Since the 1960s, with the development in theory of partial differential equations, modern mathematical tools have been developed to solve some variant coefficient equations, namely, the Fourier integral operators and pseudo-differential operator theory (Hutson and Pym, 1980; Qiu Qing-Jiu, 1985; Luan Wengui, 1989).

The role of the Fourier transform in applied mathematics for the constant coefficient linear partial differential equations is well known. Through this transform, the differential operation in the spatial—time space can be transformed into algebraic operation in the dual space; hence the Fourier transform has been successfully applied to reflection data processing. However, the Fourier transform is no longer correctly applicable to the equations with variant coefficients. As a result, mathematicians have developed some new mathematical tools, namely, the Fourier integral operator, to deal with the partial differential equation with variant coefficients.

In Section 3.1, we have mentioned some excellent properties of the constant coefficient wave equation, such as the invariance under coordinate translation, the symmetry respect to the time t, etc. But these excellent properties no longer exist if the coefficient function in the wave Eqn (3.1) becomes variant. The corresponding initial and boundary value problems of the wave equation with variant coefficients belong to generalized initial and boundary value problems. The existence of a unique solution for these problems is not guaranteed. Obviously, the initial and boundary value problem of the variant coefficient wave equation is much more complex than the corresponding problem of the constant coefficient wave equation. Notice that when the medium becomes more and more uniform, the properties of the solution of the variant coefficient wave equation must become more similar to the properties of the solution of the constant coefficient wave equation. So starting from the wave equation with constant coefficients, which is treated as a special case of the variant coefficient equations, mathematicians analyze the difficulties for the solution and invent some strategies for describing the wave field propagating in more complex medium. One of the difficulties in describing the wave field in more complex medium comes from oscillation of an exponential term in the phase function of the wave field. Therefore, mathematicians start to study the oscillatory integral. Another difficulty in describing the complex medium is that the Fourier transform cannot be implemented because the wave number is depends on the velocity function $c(x)$, so that mathematicians have to study the Fourier integral.

One may feel surprised to see that mathematicians have expanded the variant coefficient pseudo-differential operators as the superposition of a series of operators with different orders, where the base-order operator is just the constant wave differential operator. However, you may be familiar with the expansion of an ordinary function along its neighborhood by using the Taylor series. From the Taylor expansion to the pseudo-differential operator expansion, the magical tricks used by mathematicians are actually very similar. The formula in this section comes from the books by Chou, Qingjiu (1985) and Luan Wengui (1989); please use these books for more details.

6.5.2. The Oscillatory Integral

We will first discuss the oscillatory integral. Starting from the equations with constant coefficients, treated as a special case of the variant coefficient equation, we can guess the solution forms of the Cauchy problem of wave equations in general. The Cauchy problem of the wave equation with zero initial value and constant coefficients can be written as

$$
\left|
\begin{array}{l}
\dfrac{\partial^2 u}{\partial t^2} - C^2 \nabla^2 u = 0, \quad x \in \mathbf{R}^n, \quad t \in [0, +\infty] \\[2mm]
u(x,0) = 0 \\[2mm]
\dfrac{\partial u(x,0)}{\partial t} = f(x)
\end{array}
\right.
\tag{6.44}
$$

where wave field u is a complex variant function, n is an arbitrary positive integer, and x is a point in n-dimensional Euclidean space $\mathbf{R}^n$. Applying Fourier transform with respect to x on both sides of Eqn (6.44), in the wave number domain, we have a solution in terms of the wave number k:

$$
u(k,t) = (2i)^{-1} \frac{f(k)}{C|k|} \left(e^{iC|k|t} - e^{-iC|k|t} \right)
\tag{6.45}
$$

Thus the general solution of Cauchy problem Eqn (6.44) can be expressed by using inverse transform as

$$
u(x,t) = (2\pi)^{-n} \int u(k,t) e^{i\langle x,k\rangle} dk
$$

where $<,>$ represents the inner product of variants in n-dimensional Euclidean space $\mathbf{R}^n$. Substituting the above formula into Eqn (6.45), we have

$$
\begin{aligned}
u(x,t) = {} & (2\pi)^{-n} \iint e^{i\langle x-y,k\rangle + C|k|t} (2iC|k|)^{-1} f(y) dy\, dk \\
& - (2\pi)^{-n} \iint e^{i\langle x-y,k\rangle - C|k|t} (2iC|k|)^{-1} f(y) dy\, dk
\end{aligned}
\tag{6.46}
$$

where $< x, k >$ represents the linear homogenous function and y belongs to the same space $\mathbf{R}^n$ and corresponds to the initial condition $f(y)$ in Eqn (6.44).

By using operator representation, the Formula (6.46) can be written as a similar form as WKBJ approximation showed in Eqn (5.53)

$$\Re f(x, t) = \iint A(x, y, k, t) e^{i\phi(x, y, k, t)} f(y) \mathrm{d}y\, \mathrm{d}k \qquad (6.47)$$

where $\Re$ is an integral operator, different from the classical integration definition. From Eqn (6.47), we may suppose that for Cauchy problem of the variant coefficient wave equation, its solution may be similar to the form Eqn (6.47), but functions ϕ and A must take complicated forms and become simpler when C reduces to constant. In fact, from the point of view of WKBJ approximately mentioned in Section 5.5, function ϕ corresponds to the phase function and A corresponds to the amplitude function. When $A = 1$ and ϕ is a linear homogenous function of x and k, the right-hand side of Eqn (6.47) becomes the Fourier transform. Therefore, when studying the general wave equation with variant coefficients, we can assume that the solution of the problem takes the form of Eqn (6.47) and then try to redetermine the phase function ϕ and amplitude function A. We will finally show that the solution given in this way differs from the solution for the homogenous medium problem only by some smooth operators. This way shows important strategies used by mathematicians for solving wave equation with variant coefficients.

On the left-hand side of Eqn (6.47), we have denoted an integral operator $\Re$, namely, the local Fourier integral operator, being discussed in this section. In addition, it can be seen from Eqn (6.47) that the phase function $\phi(x, y, k)$ after omitting parameter t can also be represented as $\phi(x - y, k)$. If ϕ takes the form of $(x - y, k)$, the Fourier integral operator $\Re$ in Eqn (6.47) becomes the so-called **pseudo-differential operator**. Of course, rather than giving the definitions of the operators, the above statements only show the connections between the Fourier integral operator and the pseudo-differential operators together with the corresponding problems that we care about. In order to deal with these new operators, we have to give some constraints to the functions ϕ and A; otherwise the integral on the right-hand side of Eqn (6.47) may not be defined by using the classical methods. To start with, let us first discuss the situation when y vanishes on the right-hand side of Eqn (6.47), which is called the oscillatory integral

$$I_\phi(Au) = \iint A(x, \theta) u(x) e^{i\phi(x, \theta)} \mathrm{d}x\, \mathrm{d}\theta \qquad (6.48)$$

where I is the integral value of the functional. We must consider in certain real function space of ϕ and A whether the oscillatory integral

Eqn (6.48) will diverge or not. If it diverges, one cannot define an extension of the Fourier transform integral, i.e. the Fourier integral operator. If the oscillatory integral does not diverge, we can use it to define the Fourier integral operator, which must tend to the ordinary Fourier transform in special case. Now, the question is how to restrict the function space of ϕ and A in order to get convergence of the oscillatory integral Eqn (6.48)?

Before adding restrictions to the phase function ϕ, let us introduce the definition of $\wp^\infty$ function in the functional analysis that is related to the manifold. It is well known that the manifold is a generalization of the curve and curved surface in R^3 space and that the manifold is a local area in n-dimensional $\mathbf{R}^n$ space. The n-dimensional manifold that has differentiable structures is called the n-dimensional differentiable manifold. Infinitely differentiable mapping of the differentiable manifold is called $\wp^\infty$ function.

In the following discussion, we will define the phase function $\phi(x, \theta)$ in Eqn (6.48), where x and θ belong to an open set in an $\mathbf{R}^n$ space, namely, $(x, \theta) \in \Omega \subset \mathbf{R}^n$. If a real-valued function ϕ satisfies the following three conditions it is called a phase function:

1. ϕ is an infinitely differentiable function, namely, $\phi\ (x,\theta) \in \wp^\infty(\Omega)$.
2. ϕ is a linear function of θ, i.e. when $t > 0$, we have $\phi(x, t\theta) = t\phi(x, \theta)$.
3. $\phi(x, \theta)$ has no critical points with respect to x and θ, i.e. $\partial\phi/\partial x$ and $\partial\phi/\partial\theta$ are not zero at the same time.

For the real phase function ϕ as defined above, whether the oscillatory integral Eqn (6.48) diverges depends greatly on the integral variant θ.

We will next define the amplitude function A in Eqn (6.48). One of the most typical type of amplitude functions belong to Symbol $S_{\rho,\delta}^m$, which is defined by using the $\wp^\infty$ function and a threshold constant named E. The Symbol corresponds to "the pseudo–Fourier transform" in "the pseudo–wave number domain", which symbolizes an extended Fourier transform from the space–time domain to the frequency–wave number domain.

Denoting two real numbers ρ and δ satisfying $\rho > 0$ and $\delta < 1$ and an arbitrary real number m, if a function $A(x,\theta)$ satisfies the following conditions, we say that $A(x,\theta)$ belongs to the Symbol $S_{\rho,\delta}^m$:

1. $A(x,\theta) \in \wp^\infty(\Omega)$,
2. For a compact set $\Re$ in Ω and with multiple indicators α, β, there exists a constant $E_{\alpha,\beta,k}$ such that when $x \in \Re$ and $\theta \in \mathbf{R}^n$, we have:

$$\left| \delta_x^\beta \partial_\theta^\alpha A(x, \theta) \right| \leq E_{\alpha,\beta,\kappa} |\theta|^{m - \rho|\alpha| + \delta|\beta|}$$

The amplitude function $A(x,\theta)$ defined this way by using the $S^m_{\rho,\delta}$ type takes finite values and can be further expanded asymptotically. Thus Formula (6.48) can be treated as a mapping from amplitude function A to $I(Au)$. As the results indicate, the oscillatory integral Eqn (6.48) converges if this mapping is continuous.

After the amplitude function and phase function are defined, the convergence theorem of oscillatory integrals can be proved (see the book by Chou, Qingjiu, et al.). Convergence theorem for oscillatory integrals is described as follows. For any given $A(x,\theta) \in S^m_{\rho,\delta}$, $\rho > 0$, $\delta < 1$, an arbitrary real number m, and a real-valued phase function ϕ, the limit

$$\lim_{\varepsilon \to 0} \iint e^{i\phi(x,\phi)} \Psi(\varepsilon\theta) A(x,\theta)u(x)\mathrm{d}x\, \mathrm{d}\theta \qquad (6.49)$$

exists, and this limit does not depend on the specific choice of Ψ, as long as $\Psi(\theta) \in \wp_0^\infty(\Omega)$ and equals to 1 when $\theta = 0$, $u \in \wp_0^\infty(\Omega)$, where $\wp_0^\infty(\Omega)$ is the function space composed of functions $\wp^\infty$ whose branches belong to a compact set in Ω.

The oscillatory integral with both variants x and y in the functions of ϕ and A should be considered now. In this case, an integral similar to Eqn (6.48) can be written as:

$$I_\phi(Au)(y) = \iint A(x,y,\theta)e^{i\phi(x,y,\theta)}u(x)\mathrm{d}x\, \mathrm{d}\theta \qquad (6.50)$$

where Ω_y represents the range of y such that for each $y \in \Omega_y$, the integral Eqn (6.50) is convergent as stated above. Especially for $y \in \Omega_y$, if $A(x,y,\theta)$ belongs to the $S^m_{\rho,\delta}$ Symbol and the phase function $\phi(x,y,\theta)$ is real, and they are continuous with respect to y, the oscillatory integral in Eqn (6.50) is also a continuous function of y.

Thus, we have shown the convergent conditions for the oscillatory integrals having a form like Eqn (6.48). We can now define the Fourier integral operator.

6.5.3. The Fourier Integral Operator

If we fix ϕ and A, the oscillatory integral Eqns (6.48)−(6.50) becomes a mapping that tends u to $I_\phi(Au)$. So, $I_\phi(Au)$ is a linear functional with respect to elements in the function space $\wp_0^\infty(\Omega)$, which is called a Fourier distribution. According to the distribution theory in functional analysis, distribution is a linear functional satisfying certain conditions and an abstract space whose elements contain distributions is called distribution space.

As mentioned above, when Ω is an open set in $\mathbf{R}^n$, with $\wp^\infty(\Omega)$ representing all the infinitely differentiable functions and $\wp_0^\infty(\Omega)$ representing the functions of $\wp^{\infty}$ whose branches are compact set in Ω, then after defining the norm, we can define a kind of the Banach space for wave field problems, namely, a complete and locally convex linear metric space (also called Frechet space). Relating with the oscillatory integral Eqns (6.48) and (6.51), if function $u(x) \in \wp(\Omega)$, then $I_\phi(Au)$ is a linear functional of all the $u(x)$, which is also called a distribution on Ω. The space composed of all the elements of $I_\phi(Au)$ is called distribution space, it is the dual space of locally convex space $\wp(\Omega)$, denoted as $\wp'(\Omega)$. Noticing that $\wp(\Omega)$ is a function space, that $\wp'(\Omega)$ is the distribution space of the corresponding functional, and that there are important differences between the two, the Fourier integral operator can be defined as follows:

Definition: let Ω_x and Ω_y be the open sets in space $R^n{}_x$ and $R^n{}_y$, respectively; $\Phi(x, y, \theta)$ be a real-valued phase function of $x \in \Omega_x$ and $y \in \Omega_y$, and $A(x, y, \theta) \in S_{\rho,\delta}^m$ where $\rho > 0$; and $\delta < 1$. For any function $u(y) \in \wp_0^\infty(\Omega_y)$, the Fourier integral operator $\mathscr{F}$ is defined as

$$\mathscr{F}u(x) = \frac{1}{(2\pi)^n} \int_{\Re n} \int_{\Omega y} e^{i\phi(x,y,\theta)} A(x, y, \theta) u(y) dy \, d\theta \tag{6.51}$$

Corresponding to the distribution space of the functional

$$\mathscr{F}u \in \wp'(\Omega_x)$$

Therefore, the linear operator $\mathscr{F}$ maps the function space $\wp_0^\infty(\Omega_y)$ to the distribution space $\wp'(\Omega_x)$. The kernel corresponding to the Fourier integral operator is $K_F \in \wp'(\Omega_x \times \Omega_x)$. Let f denote an arbitrary function defined on Ω; the inner product between it and the kernel is

$$\langle K_F, f(x,y) \rangle \geq \int\int\int e^{i\phi(x,y,\phi)} A(x, y, \theta) f(x, y) dx \, dy \, d\theta \tag{6.52}$$

Comparing Eqn (6.52) with Eqn (6.46), one can see that this kernel is a Fourier distribution, which is a kind of extension of the Fourier integration as we expected.

The Fourier integral operator is called a pseudo-differential operator when its phase function $\Phi(x,y,\theta)$ becomes a linear homogenous function $< x - y, \theta >$. From Eqn (6.51), we have

$$\mathscr{F}_s u(x) = \frac{1}{(2\pi)^n} \int\int e^{<x-y,\theta>} A(x, y, \theta) u(y) dy \, d\theta \tag{6.53}$$

where $A \in S_{\rho,\delta}^m(\Omega_x \times \Omega_y \times \mathbf{R}^n)$. Because the phase function is limited to the linear form $< x - y, \theta >$, the dimensions of domains Ω_x and Ω_y must be the same, equal to the variant number of θ. Usually, for the pseudo-differential operator, we always consider the case of $\Omega_x = \Omega_y$. Historically, pseudo-differential operator appeared earlier than the Fourier integral operator and is believed to be a generalization of differential operators, because after Fourier transform, the linear partial differential operator could be treated as a pseudo-differential operator. Later it was found that the pseudo-differential operator was consistent with the operator defined by the oscillatory integral and had a form similar to Eqn (6.52). That is why it is called pseudo-differential operator, but, in fact, it belongs to a special kind of the Fourier integral operator.

Best of all, the linear partial differential operator with a variant coefficient can be represented by the pseudo-differential operator by using Fourier transform. Come back to the acoustic wave equation. The homogeneous wave equation can be written as

$$\mathscr{T} u = \left[\nabla^2 - N^2(x) \frac{\partial^2}{\partial t^2} \right] u(x, t) = 0 \tag{6.54}$$

where $N(x)$ is the reciprocal of wave velocity of variant, $x = (x_1, x_2, x_3)$, $k = (k_1, k_2, k_3)$, and $\mathscr{T}$ is the partial differential operator of variant coefficient corresponding to wave Eqn (6.54). Let $n = 3$ be the space dimension and t be a parameter; after applying the Fourier transform, we have

$$u(x) = \frac{1}{(2\pi)^3} \int e^{i\langle x, k \rangle} \widehat{u}(k) \mathrm{d}^3 k$$

and

$$\frac{\partial^2 u}{\partial x^2} \to (ik)^2 \widehat{u}, \frac{\partial^2 u}{\partial t^2} \to (i\omega)^2 \widehat{u}$$

So,

$$\mathscr{T} u = \frac{1}{(2\pi)^3} \int e^{i\langle x, k \rangle} \left[(ik)^2 \widehat{u} - (i\omega)^2 \widehat{u} \right] \mathrm{d}^3 k$$

Let $\widehat{u}(k) = \int e^{-i\langle x, k \rangle} u(y) \mathrm{d}^3 y$ and substituting it into the above equation, we have

$$\mathscr{T} u = \frac{1}{(2\pi)^3} \int e^{i\langle x - y, k \rangle} A(x, k) u(y) \mathrm{d}^3 y \, \mathrm{d}^3 k \tag{6.55}$$

where

$$A(x, k) = \omega^2 N^2(x) - k^2 \tag{6.56}$$

Comparing the definitions of the pseudo-differential operators in Eqns (6.55) and (6.53), we see that the partial differential operators of variant coefficient in wave equation can be represented as a pseudo-differential operator, i.e. a special Fourier integral operator.

Since the partial differential operators on the left-hand side of wave equations with variant coefficients can be represented as the oscillatory integral form as Eqn (6.55), it means that the pseudo-differential operators play an important role in solving hyperbolic equations. In fact, to solve these kinds of direct problems, we can first assume that these problems have a solution of form like Eqn (6.53) or Eqn (6.55) and then determine the phase function Φ and amplitude function A by using the Fourier integral operator theory and boundary conditions. Finally, we say that this solution is just an approximate solution to the original problem and only differs by omitting a smooth operator. In Section 5.5, we have already discussed the determination of the phase function Φ by using the eikonal equations and the amplitude function A by using transport equations, producing a high-frequency approximate solution that differs from the accurate solution by omitting some smooth operators.

6.5.4. Decomposition of Fourier Integral Operator

For the inverse problem of wave field, the generalized Radon transform is a more useful mathematical tool than Fourier transform and it has close relation with the Fourier integral operator. The generalized Radon transform of wave field equals the wave field integral over straight lines. In order to use the generalized Radon transform to represent the Fourier integral operator, let the amplitude function and phase function in Eqn (6.51) be

$$A(x, y, \theta) = \alpha(x, \theta)d(y, \theta)U(|\theta|)$$
$$\Phi(x, y, \theta) = \phi(x, \theta) - \phi(y, \theta)$$

(6.57)

respectively, where U is a real dual function that is infinitely differentiable, which can be selected in a way such that $A(x, y, \theta)$ belongs to Symbol of the type $S^m_{\rho,\delta}(\Omega_x \times \Omega_y \times \mathbf{R}^n)$.

In the spherical coordinate, $d\theta = r^{n-1}dr\, d\omega$ in Eqns (6.57) and (6.51) can be rewritten as

$$\mathcal{F}u(y) = \int_{|\omega|=1} G(y, \omega)d\omega$$

(6.58)

where

$$G(y, \omega) = \frac{1}{(2\pi)^n} \int_0^\infty \left[\int_\Omega e^{i\phi(x,y,r,\omega)} A(x, y, r, \omega)u(x)dx \right] |r|^{n-1}dr$$

(6.59)

Because of the fact that the integral of $G(y,\omega)$ in Eqn (6.58) is performed on the sphere surface $|\omega| = 1$, we only need to consider even components, which is

$$G(y, \omega) = \frac{1}{2}[G(y, \omega) + G(y, -\omega)]$$

$$= \frac{1}{2(2\pi)^n} \int_{-\infty}^{\infty} \left[\int_{\Omega} e^{i\phi(x,y,r,\omega)} A(x,y,r,\omega)u(x)dx \right] |r|^{n-1}dr \quad (6.60)$$

Using the definition of the generalized Radon transform and letting $s = \phi(x,\omega)dx$, we have

$$G'(y, \omega) = \frac{1}{2(2\pi)^n} \int_{-\infty}^{\infty} \left\{ e^{ir\Phi(y,\omega)} \, b(y, \omega)U(r) \int_{-\infty}^{\infty} e^{irs}[\mathscr{R}_a u](s, \omega)ds \right\} |r|^{n-1}dr$$

where $b(y,\omega)$ is represented as the kernel of the inverse operator of the generalized Radon transform (Yang, Wencai, 1989, pp. 177–180) and $\mathscr{R}_a$ is the generalized Radon transform. Substituting it into Eqn (6.58) and noticing

$$\int_0^{\infty} \delta[t - \phi(x, \theta)]e^{ikt}dt = e^{ik\phi(x,\theta)}$$

we have

$$\mathscr{T}u(y) = \int_{|\omega|=1} d\omega b(y, \omega) \left\{ \int_{-\infty}^{\infty} [\mathscr{R}_a u](s, \omega)k(s - s')\big|_{s'=\phi(y,\omega)}ds \right\} \quad (6.61)$$

where the integral kernel is

$$k(s) = \frac{1}{2(2\pi)^n} \int_{-\infty}^{\infty} |r|^{n-1}U(r)e^{irs}dr \quad (6.62)$$

Let $\mathscr{R}_a^*$ be the dual transform of the generalized Radon transform $\mathscr{R}_a$ and $\mathscr{K}$ be the generalized kernel operator $\mathscr{K}(s - s')$ defined by Eqn (6.62), then the Fourier integral operator defined by Eqn (6.51) can be decomposed as

$$\mathscr{T} = \mathscr{R}_a^* \, \mathscr{K} \, \mathscr{R}_a \quad (6.63)$$

So there comes the Theorem 1.

Theorem 1: the Fourier integral operator $\mathscr{T}$ defined by Eqn (6.51) can be decomposed by using Eqn (6.63), where $\mathscr{R}_a$ is the generalized Radon transform, $\mathscr{R}_a^*$ is its dual transform, and $\mathscr{K}$ is the operator whose kernel is shown in Eqn (6.62). Thus we have

$$\mathscr{T}u = \mathscr{R}_a^* \mathscr{K} \mathscr{R}_a u = \mathscr{R}_a^* \mathscr{K} v \tag{6.64}$$

where $v = \mathscr{R}_a u$ is the generalized Radon transform of the wave field and is also called the projection function and v is equal to the integration of wave field along ray path. The problem of acoustic tomography is to derive acoustic velocity variation by given the projection function v.

Theorem 1 shows the relationship between the Fourier integral operator and the generalized Radon transform. But the uncertainty of the function $b(y,\omega)$ and $k(s)$ must be noticed. Let $\mathscr{T}$ to be the pseudo-differential operator of the homogeneous wave equation with variant coefficients defined by Eqn (6.54); adding additional restrictions to the amplitude function and phase function by using asymptotic expansion, we have

$$\mathscr{T} = \mathscr{T}_0 + \sum_{i=1}^{\infty} \mathscr{T}_i \tag{6.65}$$

where $\mathscr{T}_i$ is a compact operator. It gives expansion of the pseudo-differential operators.

So far, mathematicians have expanded the pseudo-differential operator for the wave equation with variant coefficients to a superposition of a series of operators with different orders. Its base-order operator is just the wave differential operator $\mathscr{T}_0$ that is right the solution for constant co-efficient acoustic wave equation. According to Formula (6.65), to deal with the boundary value problems with complex boundary conditions, we can approximate the exact solution of the wave field step by step by increasing accuracy requirement. In the book *Principles and Techniques of Applied Mathematics* written by Freidman (1965), special discussions about the spectral expansion of some operators are worth learning. Theories and acoustic wave equations discussed in this section are the base for the application of solving boundary value problems, which will be discussed in the next section.

6.6. DOWNWARD CONTINUATION OF THE SEISMIC WAVE FIELD IN INHOMOGENEOUS MEDIUM

Here we take the downward continuation of reflection seismic wave field as an example to show how to expand the differential operator as a superposition of a series of operators with different orders. As discussed in Section 6.4, homogeneous wave equation satisfies the pseudo-differential operator equation in the vertically variant media. Now let us discuss the pseudo-differential operator equation in 3D variant media

$$\Delta u - \frac{1}{c(x,y,z)^2} \frac{\partial^2 u}{\partial t^2} = 0 \tag{6.66}$$

The problem of downward continuation of seismic wave field becomes to derive $u(x,y,z,t)$, given

$$u(x,y,0,t) = f(x,y,t); u|_{t \le 0} = 0$$

where $-H \le z < 0$ and $z = 0$ is the ground surface, H is the maximum depth.

When wave velocity does not vary and $c = $ const, the partial differential operator with constant coefficients defined on the left-hand side of Eqn (6.66) can be factorized into (Luan Wengui, 1989)

$$\Delta - \frac{1}{c^2} \frac{\partial^2}{\partial t^2} = \left(\frac{\partial}{\partial z} - i\sqrt{\frac{\partial^2}{\partial x^2} + \frac{\partial^2}{\partial y^2} - \frac{1}{c^2} \frac{\partial^2}{\partial t^2}} \right) \left(\frac{\partial}{\partial z} + i\sqrt{\frac{\partial^2}{\partial x^2} + \frac{\partial^2}{\partial y^2} - \frac{1}{c^2} \frac{\partial^2}{\partial t^2}} \right)$$

$$\tag{6.67}$$

The first term on the right-hand side of the above formula corresponds to the operator of order-reduced upgoing waves, and the second term corresponds to the operator of order-reduced downgoing waves. The lower dimensional cases have already been discussed in Section 6.1.

In the media with variant velocity, the left-hand side of Eqn (6.66) defines a pseudo-differential operator, and the upgoing wave continuation problem can be written as:

$$\frac{\partial u}{\partial z} - P\left(\frac{\partial}{\partial x}, \frac{\partial}{\partial y}, \frac{\partial}{\partial t} \right) u = 0, \quad -H \le z < 0 \tag{6.68}$$

where u is the upgoing wave. From the first term on the right-hand side of Eqn (6.67), one can speculate that the corresponding pseudo-differential operator for this problem should be

$$P = P\left(\frac{\partial}{\partial x}, \frac{\partial}{\partial y}, \frac{\partial}{\partial t} \right) = i\sqrt{\frac{\partial^2}{\partial x^2} + \frac{\partial^2}{\partial y^2} - \frac{1}{c^2(x,y,z)} \frac{\partial^2}{\partial t^2}} \tag{6.69}$$

and the initial and boundary conditions become

$$u|_{z=0} = f(x,y,t), \quad u|_{t \le 0} = 0 \tag{6.70}$$

where f is the reflected upgoing wave field received from the ground, which does not include the reflection wave generated by the ground surface.

Because Eqn (6.68) contains the pseudo-differential operator P represented in Eqn (6.69), it is better to represent P by the expansion of the partial differential operators as shown in Eqn (6.65). Therefore, considering the dual equation of Eqn (6.68) in the frequency–wave number domain

$$\frac{\partial u}{\partial z} - P(i\xi, i\eta, i\omega)u = 0, \quad -H \le z < 0 \tag{6.71}$$

where P is the Symbol of the pseudo-differential operator, corresponding to the "pseudo-wave number domain"

$$P(i\xi, i\eta, i\omega) = -\sqrt{\xi^2 + \eta^2 - \frac{\omega^2}{c^2(x,y,z)}} = -\frac{i\omega}{c}\sqrt{1 - \frac{c^2(\xi^2 + \eta^2)}{\omega^2}} \tag{6.72}$$

Notice that only a localized dual relation exists

$$(i\xi, i\eta, i\omega) \overset{\cdots}{\Leftrightarrow} \left(\frac{\partial}{\partial x}, \frac{\partial}{\partial y}, \frac{\partial}{\partial t}\right)$$

which is not a deterministic (one-to-one) relation. Even so, after substituting Eqn (6.72) into Eqn (6.71), the pseudo-differential Eqn (6.68) turns into the following equation that can be computed through the series expansion

$$\frac{\partial u}{\partial z} + \frac{i\omega}{c}\sqrt{1 - \frac{c^2(\xi^2 + \eta^2)}{\omega^2}}u = 0, \quad -H \le z < 0 \tag{6.73}$$

As the variant coefficient in the pseudo-differential operator becomes a root form, we can expand the root form as the superposition of a series of operators with different orders. However, the integral condition Eqn (6.29) in the wave number frequency domain must be considered in reflection applications. The condition required by the upgoing wave is that the vertical reflection area should be located within the "quasi circular cone"

$$\xi^2 + \eta^2 < \omega^2/c^2_{(x,y,z)} \tag{6.74}$$

which is an extension of Eqns (6.23a) and (6.29) for variant velocity. The reflected seismic wave field keeps acute angles between rays at the reflector and from the source and the receiver, where the middle line of the angle is along the z-axis. Therefore, the seismic reflection is usually wave field-consistent with the integral conditions of the "quasi circular cone".

Now expanding the root form in Eqn (6.72) by using Taylor series, the pseudo-differential operator becomes

$$P = -\frac{i\omega}{c}\left(1 - \frac{c^2}{2}\frac{(\xi^2 + \eta^2)}{\omega^2} - \frac{c^4}{8}\frac{(\xi^2 + \eta^2)^2}{\omega^4} - \frac{c^6}{16}\frac{(\xi^2 + \eta^2)^3}{\omega^6} + \cdots\right) \tag{6.75}$$

According to Eqns (6.73)–(6.75), the solution of the wave field continuation can be approximated according to accuracy requirements. Taking the first term of Eqn (6.75) as the first-order approximation and multiplying $(i\omega)$ on both sides, we have the first-order approximation equation of the pseudo-differential Eqn (6.68) in the spatial domain (x,y,t) as

$$\frac{\partial u}{\partial z} + \frac{1}{c}\frac{\partial u}{\partial t} = 0 \tag{6.76}$$

Taking the first two terms of Eqn (6.75) as the second-order approximation and multiplying $(i\omega)^3$ on both sides, we have the second-order approximation equation of the pseudo-differential Eqn (6.68) in the spatial domain (x,y,t) as

$$\frac{\partial^2 u}{\partial t \partial z} + \frac{1}{c}\frac{\partial^2 u}{\partial t^2} - \frac{c}{2}\left(\frac{\partial^2 u}{\partial x^2} + \frac{\partial^2 u}{\partial y^2}\right) = 0 \tag{6.77}$$

Similarly, the third-order approximation equation of the pseudo-differential Eqn (6.68) is

$$\frac{\partial^4 u}{\partial t^3 \partial z} + \frac{1}{c}\frac{\partial^4 u}{\partial t^4} - \frac{c}{2}\left(\frac{\partial^4 u}{\partial t^2 \partial x^2} + \frac{\partial^4 u}{\partial t^2 \partial y^2}\right) - \frac{c^3}{8}\left(\frac{\partial^4 u}{\partial x^4} + \frac{\partial^4 u}{\partial y^4} + 2\frac{\partial^4 u}{\partial x^2 \partial y^2}\right) = 0 \tag{6.78}$$

Notice that the wave velocity c in the above equations is a spatial function $c(x,y,z)$.

When using the Taylor series to expand the rational expansion Eqn (6.73), the convergence speed is not high, so we can also use other rational expansion methods, such as Pade expansion

$$\sqrt{1-x^2} \approx 1 - \frac{x^2}{2-x^2/2} - \frac{1-3x^2/4}{1-x^2/4} \tag{6.79}$$

Similarly, we can prove that the first-order approximation and the second-order approximation of the pseudo-differential Eqn (6.68) are also Eqns (6.76) and (6.77), respectively. And the third-order approximation of the pseudo-differential Eqn (6.68) by the Pade expansion is

$$\frac{\partial^3 u}{\partial t^2 \partial z} - \frac{c^2}{4}\left(\frac{\partial^3 u}{\partial z \partial x^2} + \frac{\partial^3 u}{\partial z \partial y^2}\right) + \frac{1}{c}\frac{\partial^3 u}{\partial t^3} - \frac{3c}{4}\left(\frac{\partial^3 u}{\partial t \partial x^2} + \frac{\partial^3 u}{\partial t \partial y^2}\right) = 0 \tag{6.80}$$

The above discussion shows that the variant coefficient wave operator can be decomposed by using techniques of the pseudo-differential operators. The global pseudo-differential operator for the wave field can be simplified locally to higher order partial differential operators. The constructed algorithms for finding the better solutions of downward continuation may differ only by adding more smooth operator.

6.7. DECOMPOSITION OF BODY WAVES IN REFLECTION SEISMIC WAVE FIELD

As mentioned in Section 6.5, mathematicians expand the pseudo-differential operator for the wave equation with variant coefficients to a superposition of a series of operators with different orders. Its base-order operator is just the wave differential operator $\mathscr{T}_0$ that is the correct solution for constant coefficient acoustic wave equation. According to Formula (6.65), one can approximate the accurate solution of the wave field problem with the required scale, in order to deal with some complicated boundary value problems. Looking at different waves reflected from underground reflectors, the author suggests an approach to decompose the wave field based on the idea related to Formula (6.65). The approach turns the wave field into a superposition of a series of body-wave phases with different scales, trying to reveal the relationship between the seismic phases and the geometry of reflectors.

Assuming variant velocity function and a single-point excitation, the acoustic wave equation takes the form

$$\nabla^2 u - \frac{1}{C^2(\mathbf{x})} \frac{\partial^2 u}{\partial t^2} = -\delta(\mathbf{x} - \mathbf{s}), \tag{6.81}$$

For inverse problems, one gives some observation data $u(\mathbf{x},t)$ on boundary values of the wave field and computes the velocity function $C(\mathbf{x})$; here $\mathbf{s}$ denotes the source spatial coordinates and t_0 the excitation moment for the pulse source. As mentioned in Section 6.5, Eqn (6.81) is a pseudo-differential equation, in which the pseudo-differential operator can be decomposed.

Let α_k be the disturbance of wave slowness that corresponds to local variation of the underground velocity $C(\mathbf{x})$ and k represent the local scales. The global slowly changed background of the velocity cab be denoted by $C_0(\mathbf{x})$. Under the condition

$$C(\mathbf{x}) = C_0(\mathbf{x}) \quad \text{if} \quad |\mathbf{x}| \to \infty.$$

we have

$$\frac{1}{C^2(\mathbf{x})} = \frac{1}{C_0^2} \sum_{k=0}^{\infty} \alpha_k(\mathbf{x}) = \frac{1}{C_0^2} \left(1 + \sum_{k=1}^{\infty} \alpha_k(\mathbf{x}) \right) \tag{6.82}$$

where the disturbances of the wave slowness $\alpha_0 = 1$, $\alpha_k(k > 0)$ satisfies

$$|\alpha_{k+1}(\mathbf{x})| \leq |\alpha_k(\mathbf{x})| \leq |\alpha_{k-1}(\mathbf{x})| \tag{6.83}$$

Correspondingly, we can decompose the reflection wave field into different wave phases u_k with different scales and energy; thus the reflection wave field

$$u(\mathbf{x}, t) = u_0(\mathbf{x}, t) + \sum_{k=1}^{\infty} u_k(\mathbf{x}, t) \tag{6.84}$$

Substituting Eqns (6.82) and (6.84) into Eqn (6.81), we have

$$\sum_{k=0}^{\infty} \left\{ \nabla^2 - \frac{1}{C_0^2} \frac{\partial^2}{\partial t^2} \right\} u_k(\mathbf{x}, t) - \frac{[\alpha_1(\mathbf{x}) + \alpha_2(\mathbf{x}) + \ldots]}{C_0^2} \sum_{k=0}^{\infty} \frac{\partial^2 u_k(\mathbf{x}, t)}{\partial t^2} = -\delta(\mathbf{x} - \mathbf{s}) \tag{6.85}$$

Because the continuous variation of the velocity background C_0 does not produce new seismic phase, it affects the incident wave field $u_0(\mathbf{x},t)$ only; so we can decompose Eqn (6.85) into two as follows:

$$\left[\nabla^2 - \frac{1}{C_0^2} \frac{\partial^2}{\partial t^2} \right] u_0(\mathbf{x}, t) = -\delta(\mathbf{x} - \mathbf{s}) \tag{6.86}$$

$$\sum_{k=1}^{\infty} \left\{ \nabla^2 - \frac{1}{C_0^2} \frac{\partial^2}{\partial t^2} \right\} u_k(\mathbf{x}, t) = \frac{[\alpha_1(\mathbf{x}) + \alpha_2(\mathbf{x}) + \ldots]}{C_0^2} \sum_{k=0}^{\infty} \frac{\partial^2 u_k(\mathbf{x}, t)}{\partial t^2} \tag{6.87}$$

Assume again the large-scale reflection phase as $u_1(\mathbf{x},t)$, which is coursed by large reflectors as represented by the first-order wave slowness disturbances α_1 that corresponds to first-order variation of the underground velocity $C_1(\mathbf{x})$. Similar to Eqns (6.85) and (6.87) can be further divided into

$$\left[\nabla^2 - \frac{1}{C_0^2} \frac{\partial^2}{\partial t^2} \right] u_1(\mathbf{x}, t) = \frac{\alpha_1(\mathbf{x})}{C_0^2} \sum_{k=0}^{\infty} \frac{\partial^2 u_k(\mathbf{x}, t)}{\partial t^2} \tag{6.88}$$

$$\sum_{k=2}^{\infty} \left\{ \nabla^2 - \frac{1}{C_0^2} \frac{\partial^2}{\partial t^2} \right\} u_k(\mathbf{x}, t) = \frac{[\alpha_2(\mathbf{x}) + \alpha_3(\mathbf{x}) + \ldots]}{C_0^2} \sum_{k=0}^{\infty} \frac{\partial^2 u_k(\mathbf{x}, t)}{\partial t^2} \tag{6.89}$$

where $u_1(\mathbf{x},t)$ includes direct reflection and multiples as well and their definition domain is restricted by Eqn (6.29). Logically, one can decompose Eqn (6.89) for the second-order slowness disturbances α_2 and wave field $u_2(\mathbf{x},t)$ Eqn (6.89), and so on

$$\left[\nabla^2 - \frac{1}{C_0^2}\frac{\partial^2}{\partial t^2}\right]u_2(\mathbf{x},t) = \frac{\alpha_2(\mathbf{x})}{C_0^2}\sum_{k=0}^{\infty}\frac{\partial^2 u_k(\mathbf{x},t)}{\partial t^2} \tag{6.90}$$

$$\sum_{k=3}^{\infty}\left\{\nabla^2 - \frac{1}{C_0^2}\frac{\partial^2}{\partial t^2}\right\}u_k(\mathbf{x},t) = \frac{[\alpha_3(\mathbf{x}) + \alpha_4(\mathbf{x}) + ...]}{C_0^2}\sum_{k=0}^{\infty}\frac{\partial^2 u_k(\mathbf{x},t)}{\partial t^2} \tag{6.91}$$

where u_2 represents diffraction waves, while $\alpha_2(\mathbf{x})$ represents the diffractors as discussed in Section 5.2. Letting

$$u_{\text{mic}}(\mathbf{x},t) = \sum_{k=3}^{\infty}u_k(\mathbf{x},t); \quad \alpha_{\text{mic}}(\mathbf{x}) = \sum_{k=3}^{\infty}\alpha_k(\mathbf{x}) \tag{6.92}$$

we rewrite Eqn (6.91) as

$$\left[\nabla^2 - \frac{1}{C_0^2}\frac{\partial^2}{\partial t^2}\right]u_{\text{mic}}(\mathbf{x},t) = \frac{\alpha_{\text{mic}}(\mathbf{x})}{C_0^2}\sum_{k=0}^{\infty}\frac{\partial^2 u_k(\mathbf{x},t)}{\partial t^2} \tag{6.93}$$

where, $u_{\text{mic}}(\mathbf{x},t)$ is called the microscale seismic wave field, and $\alpha_{\text{mic}}(\mathbf{x})$ is called the microscale slowness disturbances, representing some microscale velocity changes. The underground sharp velocity changes may produce microscale seismic diffraction wave field, after excitation of the incident wave.

Applying the Fourier transform with respect to t, we can obtain corresponding equations in the frequency domain. The acoustic wave equation is

$$\left[\nabla^2 + \frac{\omega^2}{C^2(\mathbf{x})}\right]u(\mathbf{x},\omega) = -\delta(\mathbf{x} - \mathbf{s}) \tag{6.94}$$

Substituting Eqns (6.82) and (6.84) into Eqn (6.94), one gets

$$\sum_{k=0}^{\infty}\left\{\nabla^2 + \frac{\omega^2}{C_0^2}\right\}u_k(\mathbf{x},\omega) + \frac{\omega^2[\alpha_1(\mathbf{x}) + \alpha_2(\mathbf{x}) + ...]}{C_0^2} = -\delta(\mathbf{x} - \mathbf{s}) \tag{6.95}$$

Because the continuous variation of the velocity background C_0 does not produce new seismic phase, we can decompose Eqn (6.95) into two as follows:

$$\left[\nabla^2 + \frac{\omega^2}{C_0^2}\right]u_0(\mathbf{x},\omega) = -\delta(\mathbf{x} - \mathbf{s}) \tag{6.96}$$

$$\sum_{k=1}^{\infty}\left\{\nabla^2 + \frac{\omega^2}{C_0^2}\right\}u_k(\mathbf{x}, \omega) = -\frac{\omega^2[\alpha_1(\mathbf{x}) + \alpha_2(\mathbf{x}) + \ldots]}{C_0^2}\sum_{k=0}^{\infty}u_k(\mathbf{x}, \omega) \qquad (6.97)$$

where $u_1(\mathbf{x},\omega)$ includes direct reflection and multiples as well and their definition domain is restricted by Eqn (6.29). Logically, one can decompose Eqn (6.97) for higher order slowness disturbances and wave field as

$$\left[\nabla^2 + \frac{\omega^2}{C_0^2}\right]u_1(\mathbf{x}, \omega) = -\frac{\omega^2\alpha_1(\mathbf{x})}{C_0^2}\sum_{k=0}^{\infty}u_k(\mathbf{x}, \omega) \qquad (6.98)$$

$$\sum_{k=2}^{\infty}\left\{\nabla^2 + \frac{\omega^2}{C_0^2}\right\}u_k(\mathbf{x}, \omega) = \frac{[\alpha_2(\mathbf{x}) + \alpha_3(\mathbf{x}) + \ldots]}{C_0^2}\sum_{k=0}^{\infty}u_k(\mathbf{x}, \omega) \qquad (6.99)$$

where $u_1(\mathbf{x},\omega)$ includes direct reflection and multiples as well. Denote $u_{11}(\mathbf{x},\omega)$ as the direct reflection, $u_{12}(\mathbf{x},\omega)$ as the second-order multiple reflection, and $u_{13}(\mathbf{x},\omega)$ as multiples higher than the second, then Eqn (6.98) can be further decomposed as

$$\left[\nabla^2 + \frac{\omega^2}{C_0^2}\right]u_{11}(\mathbf{x}, \omega) = -\frac{\omega^2\alpha_1(\mathbf{x})}{C_0^2}u_0(\mathbf{x}, \omega) \qquad (6.100)$$

$$\left[\nabla^2 + \frac{\omega^2}{C_0^2}\right]u_{12}(\mathbf{x}, \omega) = -\frac{\omega^2\alpha_1(\mathbf{x})}{C_0^2}u_{11}(\mathbf{x}, \omega) \qquad (6.101)$$

$$\left[\nabla^2 + \frac{\omega^2}{C_0^2}\right]u_{13}(\mathbf{x}, \omega) = -\frac{\omega^2\alpha_1(\mathbf{x})}{C_0^2}\sum_{k=2}^{\infty}u_{1k}(\mathbf{x}, \omega) \qquad (6.102)$$

As the diffraction waves u_2 attenuate faster than the reflection waves u_2, we denote the microscale slowness disturbances as $\alpha_{\text{mic}}(\mathbf{x})$, and rewrite Eqn (6.99) according to the above-mentioned approach

$$\left[\nabla^2 + \frac{\omega^2}{C_0^2}\right]u_2(\mathbf{x}, \omega) = -\frac{\omega^2\alpha_2(\mathbf{x})}{C_0^2}\sum_{k=0}^{\infty}u_k(\mathbf{x}, \omega) \qquad (6.103)$$

$$\sum_{k=3}^{\infty}\left\{\nabla^2 + \frac{\omega^2}{C_0^2}\right\}u_k(\mathbf{x}, \omega) = -\frac{\omega^2[\alpha_3(\mathbf{x}) + \alpha_4(\mathbf{x}) + \ldots]}{C_0^2}\sum_{k=0}^{\infty}u_k(\mathbf{x}, \omega) \qquad (6.104)$$

Let

$$u_{\text{mic}}(\mathbf{x}, \omega) = \sum_{k=3}^{\infty}u_k(\mathbf{x}, \omega); \quad \alpha_{\text{mic}}(\mathbf{x}) = \sum_{k=3}^{\infty}\alpha_k(\mathbf{x}) \qquad (6.105)$$

Similar to what we did in the time domain, one can rewrite Eqn (6.104) to the form as

$$\left[\nabla^2 + \frac{\omega^2}{C_0^2}\right] u_{\mathrm{mic}}(\mathbf{x}, \omega) = -\frac{\omega^2 \alpha_{\mathrm{mic}}(\mathbf{x})}{C_0^2} \sum_{k=0}^{\infty} u_k(\mathbf{x}, t) \qquad (6.106)$$

where $u_{\mathrm{mic}}(\mathbf{x}, \omega)$ is the microscale reflection, and $\alpha_{\mathrm{mic}}(\mathbf{x})$ is the microscale slowness disturbances.

6.8. BRIEF SUMMARY

In this chapter, we have first discussed theoretically the separation of the upward waves and the downward waves, which has different answers. For the gathered data observed also in a well, the upgoing and downgoing waves can be easily distinguished according to the sign of an event. For the data gathered with excited and observed on the ground surface, the upward and downward waves can be distinguished in the frequency—wave number domain by the difference of the wave shapes (see Formula (5.38) and elastic waves in layered media discussed in Section 3.3). However, using the factorization Eqn (6.67) of wave equations, the order-reduced upgoing and downgoing waves are different and can be defined accurately. This kind of wave decomposition has potential applications in seismic reflection.

There are two reasons for discussing the wave field downward continuation problem in this chapter. First, it is not only closely related with seismic migration but also can provide an independent tool for data processing. Second, it is a good example to demonstrate the decomposition of the pseudo-differential wave field operator, which can serve as the basis for the wave equation inversion in the next chapter. In Section 6.5, we have discussed the decomposition of the pseudo-differential operators. Only after the decomposition of the pseudo-differential operators of the wave field, the nature of the wave field propagation in variant velocity media can be revealed, and the inverse problem of wave equations has solid bases.

The decomposition theory of the pseudo-differential operators of the wave field shows that the pseudo-differential operators in variant coefficient wave equation can be expanded to a superposition of a series of operators with different orders, and the higher order operators are usually ignored in mathematical models. Then the question is what kinds of components in the wave field are ignored in the expansion method. These components seem related with the higher order singular terms in the expansion, i.e. the microscale components as mentioned in Section 6.7. The author suggests that the extraction of microscale singularity information of the wave field has a wide application in petroleum explorations (Yang Wencai and Yu ChangQing, 2008, 2009).

CHAPTER

7

Seismic Inversion

In the previous six chapters, we have briefly discussed wave propagation and the boundary value problems of wave equations, which aim to describe natural rules by following ordinary thinking, and these are called forward problems. The methods for solving forward problems mainly involve direct deductions. In this book, these methods mean correctly proposing the initial-boundary value problems related to wave equations and to given geophysical models and solving them to find the wave field propagation process. On the other hand, inversion entails inverse deductions to solve the inverse problems that will be discussed in this chapter. Inversion in this book pertains to the construction of earth models or defining the initial-boundary conditions related to wave equations by using seismic wave field data observed in limited time and space.

7.1. INTRODUCTION TO INVERSE PROBLEMS IN SEISMOLOGY

7.1.1. Inverse Problems in Seismic Exploration

Mathematically the forward problem of wave propagation can be expressed as follows: Denote a differential operator A based on physical laws and give a set of geophysical parameters M, and calculate all possible data sets D such that the following equations are satisfied:

$$A_M D = F; \quad BD = D_0 \tag{7.1}$$

where F is the given force source function, B is the operator that describes the effect of initial-boundary value conditions, and D_0 is the initial-boundary value of a wave field. In Eqn (7.1), A_M becomes a wave propagation operator suitable for the earth geophysical model M and $BD = D_0$ defines the initial-boundary value conditions to the forward problem.

In the definition of Eqn (7.1), one must consider the following aspects to construct a forward problem properly.

1. Accurate definition of D and M. Presently, D and M are primarily defined as functions or vectors in Hilbert space. The Hilbert space of M is called the model space and that of D is called the data space.
2. Selection of M and A, and matching them with B. The types of A depend on wave equations, and the types of M depend on model complexity and scales. Their properties must match the related initial-boundary conditions $BD = D_0$ properly and accurately.
3. Evaluation of the well-posed degrees of forward problems. Hypothesis is not accepted by mathematicians. They do not like to deal with problems that cannot be solved exactly. So, they observe the existence, uniqueness, and calculation stability for the solutions before solving problems. The existence of the solution to a forward problem proposed based on physical experiments can be usually proved. If the solution is absent, some logical errors may occur while converting physical models into forward problems. If the solution exists but is not unique, it is most likely that the forward problem is not well defined or that the conditions are insufficient. Assumptions must be used for turning real

physical models to corresponding mathematical problems, but completeness or redundancy of the assumptions should be exact, not more and not less. Too many assumptions may cause some contradiction and nonexistence of solutions, while too few may cause nonuniqueness. Well-posed forward problems should have solutions in given finite areas except at a few singular points. As expressed in Section 3.1, they may have a generalized solution when A is a partial differential operator with variable coefficients. The existence and the uniqueness of the solutions to forward problems cannot be guaranteed if they are not well defined.

The well posedness of mathematical problems involves the stability of the solving process. Computation errors have to be considered in solving forward problems. The solving process can be unstable if these errors are magnified exponentially during calculation. It can be known from Formulae (6.13) to (6.14) that the downward continuation of a reflection wave field mathematically belongs to ill-posed problems. For an acceptable solution of ill-posed problems, the operators can be slightly modified to allow the calculation process stable.

The inverse problems in reflection seismology can be expressed mathematically as the estimation and the evaluation of a geophysical model M that satisfies Eqn (7.1) by the given operator A and geophysical data set D_0. The operator A in seismic inversion is related to a differential operator with spatially variable coefficients, because a simple geophysical model with a constant wave velocity and density does not need to perform inversion. The goal of inversion is to extract the information of geophysical models as much as possible from the observed seismic data D_0, and to avoid any possibility of being artificially misled. To find an accurate solution of inverse problems, the following aspects should be considered.

1. Deduction of the inverse operator A^{-1} and analysis of its attributes. For inverse problems, the operator A of forward problems is usually given already, so the construction of the corresponding inverse operator becomes the main subject of this chapter. Just as discussed in Chapter 6, partial differential operators with variable coefficients can be simplified as quasidifferential operators, and their inverse operators belonging to Fourier integral operators generally. It is possible to construct the equation for a geophysical model M only after the inverse operator is derived. Once the inverse operators are formed, the first thing to do is to examine whether the inversion problem has been defined correctly or not, that is evaluate the existence, uniqueness of the solutions and well posedness of the procedures under the assumption that the data set D_0 is accurate and sufficient.

2. Mathematical presentation of the real data sets D_o and geophysical model M, and analysis of their properties, such as dimension, sampling rate, observation scale, and accuracy. The complexity of the assumed geophysical models must match the quality and quantity of observed seismic data. For instance, single component seismic data can only match with models of acoustic velocity disturbance, but not with models with elastic parameters.

3. Different from forward problems, all seismic inversion problems only have generalized solutions, and are rarely unique. The observed data are usually not dense enough and have errors, so the corresponding solutions are not unique.

4. Criteria and algorithms for finding the generalized solution should be formed for most seismic inverse problems. Furthermore, the evaluation of a computed solution becomes important to appraise fineness of the solution among many possible solutions.

7.1.2. The Generalized Solutions

As discussed in Sections 3.1 and 6.5, the solutions of the forward problems of the wave propagation in media of a variant velocity may belong to generalized solutions. Almost all solutions of inverse problems belong to generalized solutions. Generalized solutions can be divided into different types, they are called quasisolutions, approximate solutions, and equivalent solutions.

[1] The Quasisolution

Let us assume that a data set with errors $\tilde{d} \in D$ is a real data set in the data space D, and $d = A(m)$ is the accurate data set without errors. Then the solutions of minimizing the span between $\tilde{d}$ and d in data space D are called quasisolutions with respect to the Earth model m. The existence of quasisolutions has been proved by Tikhonov and Ivanov. Quasisolutions are relevant to data errors. When the data error is too big, $\tilde{d}$ may not be located in the theoretical data space D, and the quasisolution depends on the projection of $\tilde{d}$ on a certain manifold in the data space D. If we assume that m is a compact real parameter set, $m \in M$, A is a continuous operator and p is a projection of D into $A(m)$, then there is a unique solution for a precise data set $d = A(m)$. For any real data set $\tilde{d}$, there will exist a quasisolution that continuously depends on $\tilde{d}$ as

$$X = A^{-1}\left[p(\tilde{d})\right]. \tag{7.2}$$

where X is called the quasisolution with respect to m. The existence of a quasisolution makes seismic inversion possible to perform with error-contained data. However, another question arises as to how this can be done if the data are incomplete?

[2] The approximate solution

All approximate solutions of geophysical models are located in a region of model space. Selecting one from this region according to certain criteria of minimizing some functional leads to an approximate solution. The minimization criteria of the functional are based on the computation of distance in a normed vector space D, or in a normed vector space M. For the example of linear inversion, let $x(t)$ and $y(t)$ be two elements in the normed space. Then, the minimization can be done under the following criteria (Buckas and Gilbert, 1972; Tikhonov and Arsenin, 1977; Lines and Treitel, 1983; Marozov, 1984):

a. Minimizing the super distance, $\mathrm{dis}(x, y) = \sup |x(t) - y(t)|, \ t \in [a, b]$;

b. Minimizing the second moment, $\mathrm{dis}(x, y) = \left[\int_a^b |x(t) - y(t)|^2 \mathrm{d}t \right]^{1/2}$;

c. Minimizing the weighted second moment,

$$\mathrm{dis}(x, y) = \left[\int_a^b |x(t) - y(t)|^2 w(t) \mathrm{d}t \right]^{1/2};$$

where $w(t)$ is a weight function.

If one follows the principles stated above, and assumes that x, x' are two elements in the model space, then the distance between these two elements can be written as $\mathrm{dis}_m(x, x')$ and $\mathrm{dis}_d(\tilde{d}, A(x))$ written as the distance between data sets $\tilde{d}$ and errorless data d; then, the so-called cost functional is

$$\Phi(x) = \lambda \mathrm{dis}_m(x, x') + (1 - \lambda)\mathrm{dis}_d(\tilde{d}, A(x)), \lambda \in (0, 1) \tag{7.3}$$

In the inversion theory proposed by Buckas–Gilbert (1968–1972), λ is called the compromise factor. The first item on the right side of Eqn (7.3) means the minimizing modulo of solution estimations to limit the amplifying data error, and the second term mainly means maximizing the fitness of the observed data to enhance the resolution of the solution estimations to find a good earth model. Because these two items work against each other, we have to use λ, which acts as a compromise. This factor is called the regularization factor in regularization theory (Tikhonov, 1963). The solution estimation produced by minimizing the cost functional is called approximate solution, which is an extension of the quasisolution and is commonly used in seismic inversion.

In this context, some hints about regularization theory should be explained in details. Assuming that δ is the modulo of the data errors, and

δm is the modulo of the disturbance of solution estimations related to a linear geophysical inversion problem, Lai Wengui (1989) divided the inverse problems into the following three types:

1. Well-posed type, where the modulo of the disturbance of solution estimations is proportional to that of errors, that is $\delta m \propto \kappa \delta$;
2. Power type, $\delta m \propto \delta^{\alpha}, 0 < \alpha < 1$;
3. Exponential type, $\delta m \propto \left(\ln \frac{1}{\delta}\right)^{-\alpha}$.

The third type belongs to the so-called ill-posed problems that cannot be solved mathematically by any definite approach. How can ill-posed problems be changed to solvable problems like well posed ones? The ill posedness of an inverse problem is generated by the singularity of the operator A that appears in original forward problems (Garnir, 1980). An inverse problem can turn to a well-posed one by replacing the operator A with a regularized operator, and the turning procedure is referred to as regularization. For example, in a linear inverse problem $Ax = d$, where $d \in D$, the regularization operator R should satisfy

$$\lim R(Ax, \lambda) = x, \quad \text{for } \lambda \to 0, \tag{7.4}$$

where $\lambda > 0$. This is the basic idea in the regularization theory.

[3] The equivalent solutions

Instead of minimizing the cost function Eqn (7.3), let

$$\Phi(x) \le \varepsilon, \tag{7.5}$$

Then, the solution estimations obtained by using Eqn (7.5) is called the equivalent solution. The solution estimations of geophysical inverse problems, obtained by applying the maximum likelihood criterion, the minimal entropy criterion, and the genetic algorithms, all belong to equivalent solutions.

Because the solution estimates of an inverse problem are not unique, and the procedures for solution estimates are chosen artificially, appraisal of the solution estimates would be important. Tarantola and others (1987) have developed a posterior estimation inversion approach, called the posterior probability density (PPD), by analyzing data errors based on Bayes' theorem in the probability theory. Natural events usually occur randomly. If the probability of an event A is P under some conditions, when the conditions repeat n times, the event A occurs k times, the probability of k-time occurrences of event A is denoted by $P_{n,k}$, then $P_{n,k}$ is proportional to

$$P^k (1 - P)^{n-k}, \quad (k = 0, 1, \ldots, n).$$

If the data error follows a Gaussian distribution, the estimated solution with the maximum PPD can approach the real solution. Therefore, maximizing PPD serves as a new criterion for solution estimations of geophysical inversion. The application of the criterion based on Bayes' theorem will be discussed at the end of this chapter.

7.1.3. Linearized Iterative Seismic Inversion

Since the 1980s, linear inversion theories have been applied in geophysics (Menko, 1984; Yang Wencai, 1989), and the solutions of linear inversion problems can successfully be estimated and evaluated with the help of modern computers. For nonlinear problems in geophysics, we may try to construct procedures to get an approximate solution estimate by simplifying nonlinear operators to a linear operator. As inversion of the acoustic wave equation is actually nonlinear, the Born and Rytov approximation methods for inversion have been proposed for many years, and will be discussed in the following sections.

Typical seismic inversion problems were divided into four types by Tarantola and others (Tarantola, 1987; Beylkin and Rurridge, 1990), that is linear, weakly linear, pseudolinear, and strong nonlinear. Correspondingly, these types divided based on operator properties must correlate to four different algorithms: linear inversion, generalized linear inversion, pseudolinear inversion, and nonlinear inversion. In fact, such a division is not enough for guiding seismic inversions. The nonlinearity of inverse problems depends not only on the properties of the wave propagation operator but also on the errors of input data and the complexity of geophysical models.

Linearized inversion methods are commonly used to solve nonlinear inverse problems (Sabatier, 1985; Cohen and Bleistein, 1986; Tarantola, 1987). To form a proper procedure of linearized inversion, the convergence and the rule of stopping iterations become the key factors. The author suggests that it is very important to check what kind of rules the nonlinear iteration should obey and what status the current iteration goes into. Although the state equation of inversion iterations cannot be deduced from physics directly, finding some characteristic parameters, which describe the iteration state, can be very helpful for controlling the iteration successfully. Nonlinear iterations should follow the general rules of natural nonlinear dynamic processes, in which the errors contained in initial-boundary conditions will result in an unpredictable trajectory path in solution space, and the inversion process will eventually become chaotic. The author (1993) improved the chaotic process that occurred during a seismic inversion process via numerical experiments. There are several states from stable phases to the chaotic phase in the process. How do the phases change? What kind of parameters characterize the phase changes? It has been discovered for seismic inversions that nonlinear seismic inversion starts from the phase of rapid improvement of resolution, then the phase of slow resolution improvement and sharp growth of variance. The critical phase follows with no resolution improvement, and finally, the iteration becomes chaotic. The magnitude of the data error has little effect on the critical and chaotic states, but has a great effect on the phase-change speed. Large data errors in nonlinear inversions make quick phase-changes in iterations and make the iteration procedure chaotic at a few iteration cycles (Schuster, 1987; Yang Wencai, 1993, 1997; Nie Yan-Liang et al., 1995; Kaplan and Glass, 1995). To find a good solution estimation of a high resolution and small variance, one must look at the iteration state before the process goes to the phase of sharply growing variance. In summary, while working on nonlinear seismic inversion, one should compute a characteristic parameter called the Lyapunov exponent, which shows the states of the iteration. A sudden change in the Lyapunov exponent indicates a phase change, describing the system status. A better use of the Lyapunov exponent can help us to control the iteration process and stop the iteration, instead of finding an optimal compromise between high resolution and small variance.

Another difficult problem of the linearized iteration lies in the location of solution estimates in model space, that is whether the estimate locates at the local minimum or global minimum? This is a problem for all inversion methods based on operator differentiation. In most cases, the initial model used by iterative inversion procedures will have some impact on the solution estimate or the inversion process. It is better to figure out several possible models as the initial model for iterative inversion procedures. If one has no idea about the earth velocity model, then one should try a constant velocity model as the initial model, to avoid being misled by some unpredictable facts.

7.1.4. Nonlinear Stochastic Inversions

Quasi or approximation solutions must be mathematically derived based on wave equations and the given optimization criterion to form the corresponding algorithms. These solution estimations will be discussed in detail in the following sections of this chapter. For nonlinear inversions with complicate earth models, the derivation of quasi and approximation solutions is very difficult sometimes analytically. Therefore, we have to try some new methods for geophysical inversion, and among them the equivalent solution becomes a good option. It has been proved that equivalent solutions got by applying the stochastic strategy can be significant to some degree.

Nonlinear stochastic inversion originates from the trial-and-error method in computational mathematics and it makes full use of the high-speed computational capacity of modern computers. As long as forward problems are given, the method can perform inversion directly without much logical deduction for inversion formulae. A key to nonlinear stochastic inversions is to construct an effective criterion for modifying the model parameter increment so as to reduce iteration times and to prevent divergence. In the case when forward calculation is simple and can be computed quickly, as well as the total number of model parameters is not large, stochastic inversion methods can be relevant.

The simulated annealing and genetic algorithm are often accepted in geophysical inversion, good for nonlinear inversion if effective parameters increment modifying criterion is given (Roseman, 1985; Aarts, and Korst, 1989; Davis, 1991). The simulated annealing method is the computer simulation of metal annealing studied in metallurgy. As cooling proceeds, atoms in a metal get arranged in order and approach the lowest energy state. As the lowest energy state equals the minimization of the energy, simulated annealing shows a successful example of minimization methods for solving nonlinear dynamic problems. In the process of seismic inversion, a numerical system is built between observed data and model parameters; they are related by a nonlinear operator. In the system fitting difference between real and calculated data in L_2 space acts as the "energy" of the system. An equivalent solution can be approximated iteratively by minimizing the system energy through a simulated annealing process. The process of searching in model space can be demonstrated by an analogy of finding a man in a city starting from the train station (as the initial model) and being guided by a city map; asking directions step by step equals to guiding by the operator differentiation; walking a segment equals to doing a cycle of linear iteration. It is troublesome if there are several places with the same name as that of the right location the man is going to. Instead of guiding by a "map", the simulated annealing method tries a few steps with different directions first to find which direction is closer to the target. If one repeats this procedure again and again, the target will be probably approached. However, this method has no guarantee for global convergence as it could turn into the wrong direction at some steps midway of the iteration.

In 1953, Metropolis proposed the cycle sampling algorithm that provides practical criteria for the simulated annealing method. Let V be the state of a thermodynamic system at temperature T. Then, random disturbance of the dynamic system presents a new state V' in which energy change $\Delta\phi = \phi(V') - \phi(V)$. In the new state, acceptable energy minimization follows the exponential law of probability as (Metropolis, 1953)

$$P = e^{-\Delta\phi/T} \tag{7.6}$$

In other words, if the probability of the new state computed with Eqn (7.6) becomes large, then randomly produced new model parameters are going in the right direction during the iterations analogous to the simulated annealing. The model modifying criterion in the simulated annealing inversion follows the exponential law Eqn (7.6) proposed by Metropolis. Because of the similarity between the minimization of fitting variance and the simulated annealing algorithm, the latter can be easily employed in seismic inversions.

Another natural nonlinear process relates to biological reproduction, but the genetic evolution does not follow the second law of thermodynamics. The genetic algorithm, proposed by John Holland in 1975, shows the general trend of natural random selections that obey the law of the survival of the fittest. This algorithm originates from the computer simulation of natural genetic evolution of living things (Holland, 1975; Roseman, 1985; Davis, 1991). The first step in the algorithm performs the coding of the model parameters; these codes represent gins of the initial parents of the earth model. The second step in the algorithm performs random exchanges of the codes, producing a code set representing gins possible new generation of the earth model. The third step in the algorithm performs a minor variation of the code set and breeding a new generation family for natural selection. Finally, one has to compute the probability of crossmating to control the iterations and to select the best fit for the next iteration. Application in various fields shows that the genetic algorithm is robust and can usually converge on to an optimal solution if the probability of cross-variation and breeding is properly computed. However, since seismic inversion is different from organic evolution, there are some questionable points about the procedure. For example, what is the physical meaning of the probability of crossvariation and breeding? Early maturing might occur in the practical performance of the algorithm, which means that the reproduction is so quick that the procedure is not convergent at the global optimal solutions. Moreover, the iterative modification of model estimates often becomes insensitive to the fit, so that the earth models cannot be improved significantly after a lot of computing time.

The genetic algorithm is good for multichannel computation performed in parallel computers. But its mathematical and physical foundations are not as solid as that for simulated annealing. The genetic algorithm can be improved by introducing the Metropolis sampling theorem, which works well in simulated annealing methods.

7.2. BORN APPROXIMATION INVERSION BY INVERSE SCATTERING

The inversion of acoustic Eqn (7.7) is the process of computing a wave velocity $C(\mathbf{x})$ by a given boundary data set $u(\mathbf{x},t)$ (Cohen and Bleistein, 1979, 1986):

$$\nabla^2 u - \frac{1}{C^2(\mathbf{x})} \frac{\partial^2 u}{\partial t^2} = -\delta(\mathbf{x} - \mathbf{s})\delta(t - t_0), \tag{7.7}$$

where s denotes the location of a single source, and t_0 is the moment of shooting. As discussed in Section 6.5, Eqn (7.7) is a quasidifferential equation, in which the quasidifferential operator can be decomposed. Inversion requires the inverse operator of the quasidifferential operator, so we must first decompose the inverse operator. The variable underground wave velocities can be divided into two parts: the background velocity c_0 that globally varies continuously and smoothly, and disturbance α that varies locally and sharply as

$$\frac{1}{C^2(\mathbf{x})} = \frac{1 + \alpha(\mathbf{x})}{C_0^2(\mathbf{x})}, \tag{7.8}$$

where α is called the square slowness disturbance, satisfying

$$-1 \leq |\alpha(\mathbf{x})| \leq 1 \quad \text{and} \quad C(\mathbf{x}) = C_0(\mathbf{x}) \quad \text{if } |\mathbf{x}| \to \infty. \tag{7.9}$$

The real disturbance of the wave velocity is denoted by β as follows:

$$\frac{1}{C^2(\mathbf{x})} = \frac{1}{C_0^2(1 + \beta)^2}, \quad \beta = \Delta C / C_0$$

It is easy to prove that

$$\alpha = \beta \left(-2 + 3\beta^2 - 4\beta^3 + 5\beta^4 - 6\beta^5 + \cdots \right) \tag{7.10}$$

The media of a continuously variable velocity background C_0 does not create new waves, so C_0 is relevant only to the incident wave $u_0(\mathbf{x},t)$. Locally and rapidly, variable velocities create new types of waves, so α is related to both incident waves and scatter waves. Here, we assume that reflection waves, diffraction waves, and random scattered waves all belong to scatter waves $u_s(\mathbf{x},t)$. Thus, the acoustic wave field

$$u(\mathbf{x}, t) = u_0(\mathbf{x}, t) + u_s(\mathbf{x}, t). \tag{7.11}$$

Substituting Eqns (7.11) and (7.8) into Eqn (7.7) and performing the Fourier transformation on t, we get the wave equation for the incident waves

$$\left[\nabla^2 + \frac{\omega^2}{C_0^2(\mathbf{x})} \right] u_0(\mathbf{x}, \mathbf{s}, \omega) = -\delta(\mathbf{x} - \mathbf{s}) \tag{7.12}$$

and that for the scatter waves

$$\left[\nabla^2 + \frac{\omega^2}{C_0^2(\mathbf{x})} \right] u_s(\mathbf{x}, \mathbf{s}, \omega) = -\alpha(\mathbf{x}) \frac{\omega^2}{C_0^2(\mathbf{x})} u(\mathbf{x}, \mathbf{s}, \omega). \tag{7.13}$$

Note that Eqn (7.13) is a nonlinear equation. If the scatter waves are far weaker than the incident waves, Eqn (7.13) can be rewritten with the Born approximation as

$$\left[\nabla^2 + \frac{\omega^2}{C_0^2(\mathbf{x})}\right] u_s(\mathbf{x}, \mathbf{s}, \omega) \approx -\alpha(\mathbf{x})\frac{\omega^2}{C_0^2(\mathbf{x})}u_0(\mathbf{x}, \mathbf{s}, \omega), \quad \text{if } u_s \ll u_0 \quad (7.14a)$$

This equation is linear. With the Born approximation, the ratio of scatter waves to incident waves can also be written as

$$\left[\nabla^2 + \frac{\omega^2}{C_0^2(\mathbf{x})}\right]\frac{u_s(\mathbf{x}, \mathbf{s}, \omega)}{u_0(\mathbf{x}, \mathbf{s}, \omega)} \approx -\alpha(\mathbf{x})\frac{\omega^2}{C_0^2(\mathbf{x})} \quad (7.14b)$$

The solution of Eqn (7.14) can be expressed using the Green function. According to the definition, the corresponding Green function must satisfy

$$\nabla^2 G(\mathbf{x}, \mathbf{s}, \omega) + \frac{\omega^2}{C_0^2(\mathbf{x})}G(\mathbf{x}, \mathbf{s}, \omega) = -\delta(\mathbf{x} - \mathbf{s}). \quad (7.15)$$

It is clear by comparing Eqn (7.15) with Eqn (7.1) that the Green function is equivalent to the incident waves of a single source. Assume that r denotes a receiver point; then, the scatter waves satisfy the Born approximation:

$$u_s(\mathbf{r}, \mathbf{s}, \omega) \approx \omega^2 \int \frac{G(\mathbf{x}, \mathbf{s}, \omega)G(\mathbf{x}, \mathbf{r}, \omega)}{C_0^2(\mathbf{x})}\alpha(\mathbf{x})d\mathbf{x}, \quad (7.16)$$

It is an integral equation for the inversion of slowness disturbance $\alpha(\mathbf{x})$. Mathematically, Eqn (7.16) is a Fredholm integral equation of the first kind, where $G(\mathbf{x},\mathbf{s},\omega)$ equals the incident wave of a single source and $G(\mathbf{x},\mathbf{r},\omega)$ is the virtual incident wave if the single source would be located at the receiver point.

For the extreme case of homogeneous media, $C_0(\mathbf{x}) = \text{const}$. By referring to Eqn (5.29c), we can write the incident wave and its Green function as

$$G(\mathbf{x}, \mathbf{s}, \omega) = u_0(\mathbf{x}, \mathbf{s}, \omega) = \frac{\exp(ik|\mathbf{x} - \mathbf{s}|)}{4\pi|\mathbf{x} - \mathbf{s}|}, \quad k = \omega/C_0; \quad (7.17)$$

Similarly, the virtual incident waves and its Green function are

$$G(\mathbf{x}, \mathbf{r}, \omega) = u_0(\mathbf{x}, \mathbf{r}, \omega) = \frac{\exp(ik|\mathbf{x} - \mathbf{r}|)}{4\pi|\mathbf{x} - \mathbf{r}|}. \quad (7.18)$$

By substituting Eqns (7.17) and (7.18) into Eqn (7.16), we get the linear inversion equation that can be used in the seismic inversion of prestack data.

First of all, we must describe the data sets and the earth model for seismic inversion. The earth model is represented by the square wave slowness disturbance $\alpha(\mathbf{x})$ in Eqn (7.16). Although the data set is expressed as the incident wave field u_s, it has been observed only at a limited number of receivers and depends on data acquisition gathers. The inverse scattering inversion of the Born approximation for three different data gathers will be discussed as follows:

[1] Poststack common shot-receiver gather

Assume homogeneous media and $\mathbf{s} = -\mathbf{r}$, according to Eqns (7.16)–(7.18) the inversion equation for a poststack common shot-receiver gather becomes

$$u_s(\mathbf{r}, \omega) \approx k^2 \int_{z>0} \alpha(\mathbf{x}) \frac{\exp(2ik|\mathbf{x} - \mathbf{r}|)}{(4\pi)^2 |\mathbf{x} - \mathbf{r}|^2} \, d\mathbf{x}. \tag{7.19}$$

After the following transformation proposed by Cohen and Bleistein (1979, 1986)

$$\overline{u}_s(\mathbf{r}, \omega) = -i \frac{\partial}{\partial \omega} \left[\frac{u_s(\mathbf{r}, \omega)}{\omega^2} \right] = \frac{1}{2\pi C_0^2} \int \alpha(\mathbf{x}) \frac{\exp[2i\omega|\mathbf{x} - \mathbf{r}|/C_0]}{4\pi|\mathbf{x} - \mathbf{r}|} d\mathbf{x}, \tag{7.20}$$

the scattered waves can be converted into a special convolution integral, which is in accordance with the convolution integral shown in Section 5.2. Actually, the reflection seismic trace from 1D strata with surface receivers can be represented as

$$\overline{u}_s(0, \omega) = \frac{1}{2\pi C_0^2} \int \alpha(z) \frac{\exp[2i\omega z/C_0]}{4\pi z} dz, \tag{7.21}$$

It is clear that the seismic traces of a common shot-receiver gather contain a set of time shift impulses whose amplitudes are determined by slowness disturbance, which is the theoretical foundation of the seismic deconvolution after the Born approximated inversion. In conclusion, using the seismic deconvolution procedure for inversion of reflectivity function is equivalent to the Born approximated inversion.

[2] Waveform inversion for a CMP gather in the frequency domain.

Assume that the locations of the midpoint, the shot, and receiver are $(0,0,0)$, $(-x,0,0)$, and $(x,0,0)$, respectively, and that a small bin of underground horizontal reflectors is located at $y = (0,0,h)$ with an area ds and a thickness dh (Figure 7.1). As the normal direction of the reflector points vertically, the angle between the vertical and the incident ray satisfies

$$\cos \varphi = \frac{h}{r}; \; r = \sqrt{x^2 + h^2} \tag{7.22}$$

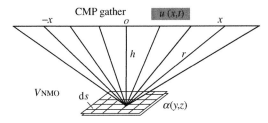

FIGURE 7.1   Geometry and notations of a CMP profile.

According to Eqn (5.29c), the Green function (7.18) can be written as follows:

$$G(x, y, \omega) = \frac{1}{2\pi r} e^{i\omega r/c_0} \tag{7.23}$$

by taking the surface effect into consideration. Putting Eqn (7.23) into Eqn (7.16) and integrating only to the disturbance source $dy = dh\,ds\,\cos\phi$, we get

$$u_s(x, \omega) = \frac{\omega^2}{(2\pi c)^2} \frac{h\,dh\,ds}{(x^2 + h^2)^{\frac{3}{2}}} \alpha(0, h) e^{i2\omega\sqrt{x^2+h^2}/c_0} \tag{7.24}$$

where $\alpha(y) = \alpha(0,h)$. The waveform inversion of the CMP gather in the frequency domain uses $\{u_s(x,\omega)\}$ as the input to get $\alpha(0,h)$ as the output. The parameter h in Eqn (7.24) can be easily calculated through the arrival time $T_0 = 2h/c_0$ at location $x = 0$.

As the input data are often of redundancies, the variations of $\alpha(0,0,h)$ along half offset $x/2$, or equivalently the ray angle φ, can be searched, providing a kind of amplitude versus offset inversion. It is easy to compute the square wave slowness disturbance $\alpha(0,h)$ by solving linear

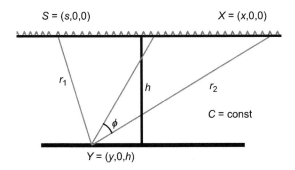

FIGURE 7.2   Geometry and notations of a common source gather and profile.

Eqn (7.24), and then converting it into the reflection coefficient by using Eqn (7.10).

[3] Waveform inversion for a common shot gather in the frequency domain

Assume that the shot, receiver, and midpoint are located at $(s,0,0)$, $(x,0,0)$, and $((x+s)/2,0,0)$, respectively, and the location of an underground reflection cell is $y = (y,0,h)$ with an area ds and thickness dh (Figure 7.2). The normal of the reflector cell surface is in the vertical direction, and the length of the incident ray is r_1 and that of the reflective ray is r_2; we have

$$r_1 = \sqrt{(s-y)^2 + h^2};\tag{7.25}$$

$$r_2 = \sqrt{(x-y)^2 + h^2}\tag{7.26}$$

The angle between the incident wave and the reflection equals 2φ and is easy to be computed by using ray tracing. The depth h can be calculated from the arrival time $T_0 = 2h/c_0$ at $x = 0$. After consideration of the surface Eqn (7.23) can be rewritten as

$$G(s,y,\omega) = (1/2\pi r_1)e^{i\omega r_1/c_0}\tag{7.27}$$

$$G(x,y,\omega) = (1/2\pi r_2)e^{i\omega r_2/c_0}\tag{7.28}$$

Substituting these two formula into Eqn (7.16) for disturbance $\alpha(y) = \alpha(y,0,h)$, and performing the integration with

$$dy = dy \, ds \, \cos \varphi(y),\tag{7.29}$$

we have

$$u_s(s,x,\omega) = \frac{ds \, \omega^2}{(2\pi c)^2} \int dy \, \frac{\cos \varphi}{r_1 r_2} \alpha(y,h)e^{i\omega(r_1+r_2)/c_0}\tag{7.30}$$

Again Eqn (7.30) is a Fredholm integral equation of the first kind, which can be used for the inversion of the square wave slowness disturbance $\alpha(y,0,h)$. Waveform inversion for a common shot point (CSP) gather in the frequency domain uses $\{u_s(x,\omega)\}$ as the input to compute $\alpha(y,0,h)$ as the output. The discrete form of Eqn (7.30) is

$$u_s(s,j,\omega) = \frac{ds \, dy \, \omega^2}{(2\pi c)^2} \sum_j \frac{\cos \varphi(j)}{r_{1j} r_{2j}} e^{i\omega(r_{1j}+r_{2j})/c_0} \alpha(j,h)\tag{7.31}$$

This equation is linear and can be solved by applying the generalized linear inversions (Yang, 1989, 1997).

7. SEISMIC INVERSION

In practice, seismic profiling contains many CSP gathers, and Eqn (7.31) can be used and put into a matrix form as

$$u = B\alpha \tag{7.31a}$$

The procedure of prestack waveform inversion for CSP gathers by using the Born approximation in the frequency domain can be constructed directly based on Eqn (7.31) as shown by a flowchart in Figure 7.3. The procedure contains following steps:

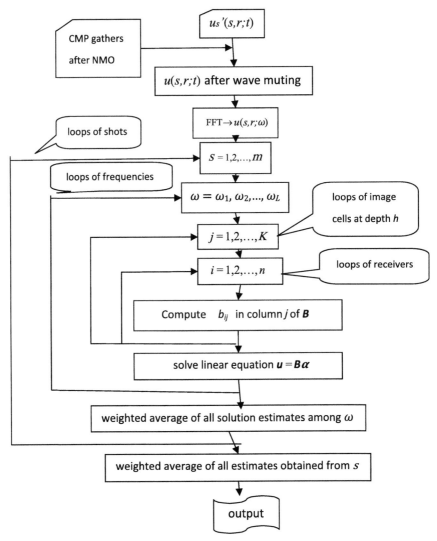

FIGURE 7.3 Logical flowchart of the waveform inversion for prestack CSP gathers in the frequency domain.

1. Input prestack CSP gather $u'(s,r,t)$ and background velocity C_o, where C_o can be replaced by the average velocity from the surface to the corresponding depths.
2. Extract the waves of target events from the CSP data, or mute irrelevant events to get wave field data set $u(s,r,t)$ for inversion.
3. Convert $u(s,r,t)$ to $u(s,r,\omega)$ in the frequency domain by Fourier transformation.
4. Repeat steps 1–3 for different CSP gathers by cycles $s = 1,2,\ldots,m$. The selection of frequencies, $\omega = \omega_1, \omega_2, \ldots$, for a single frequency inversion by using Eqn (7.31).
5. Compute the square slowness disturbance repeatedly for each cell denoted by $j = 1,2,\ldots,K$, located at a depth h.
6. Compute the element b_{ij} of the coefficient matrix B in Eqn (7.31a), where $i = 1,2,\ldots,n$. Repeat the computation to each receive point in the CSP gathers until $i = n$.
7. Solve Eqn (7.31a), $u = B\alpha$, and find the square slowness disturbance α.
8. Return to step 5, and get the average solutions with different frequencies.
9. Return to step 4, and get the average solutions with different shot gathers.
10. Convert the square slowness disturbance into reflectivity by using Eqn (7.10).

The size of the cell must be proper with the shots and receivers during the inversion.

7.3. ACOUSTIC WAVE EQUATION INVERSION IN VERTICALLY INHOMOGENEOUS BACKGROUND MEDIA

We have shown the solutions of an acoustic wave equation for media with linearly variable velocity in Section 5.4. Before studying more complicated inversion models, we should discuss its inverse problem, which can be treated as the first-order approximation of reflection seismic wave inversions. The acoustic velocity linearly variable in vertical coordinates is represented as (Huang Liang-Jiu and Yang Wencai, 1991; Yang Wencai and Du Jian-yuan, 1992, 1993)

$$C_o(z) = a + bz, \quad b > -a/z_{max}$$

where $x = (x,y,z)$, and z_{max} is the maximal reflection depth and $z < z_{max}$. The corresponding equation that describes the propagation of a harmonic wave in the frequency domain can be written as

$$\left(\nabla + \frac{\omega^2}{C^2(z)}\right)u(x,x',\omega) = -\delta(x - x') \tag{5.32}$$

In this equation, the left side is a pseudodifferential operator, and the item for initial condition at its right side is neglected, meaning that the excitation starts at $t = 0$.

Similar to the discussion in the last section about the Born approximation, except with a variable wave velocity, the underground variable velocity $C(z)$ can be divided into two parts: the linearly background velocity $C_0(z)$ of varying continuously, and velocity disturbance α varying rapidly and locally. They satisfy

$$\frac{1}{C^2(\mathbf{x})} = \frac{1 + \alpha(\mathbf{x})}{C_0^2(z)},$$

where α is the square slowness disturbance and should satisfy

$$-1 \leq |\alpha(\mathbf{x})| \leq 1 \quad \text{and} \quad C(\mathbf{x}) = C_0(z) \quad \text{if } |\mathbf{x}| \to \infty.$$

By assuming that $C_0(z)$ is continuous and differentiable, so that no scatter waves are created by the background velocity changes, we have $u = u_0 + u_s$, where the incident wave $u_0(x,t)$ is relevant only to C_0, α is related to both the incident waves and scatter waves. Substituting Eqn (7.8) into Eqn (5.32) we have the equation for both the incident waves

$$\left[\nabla^2 + \frac{\omega^2}{C_0^2(z)} \right] u_0(\mathbf{x}, \mathbf{s}, \omega) = -\delta(\mathbf{x} - \mathbf{s}). \tag{7.12a}$$

and the scatter waves,

$$\left[\nabla^2 + \frac{\omega^2}{C_0^2(z)} \right] u_s(\mathbf{x}, \mathbf{s}, \omega) = -\alpha(\mathbf{x}) \frac{\omega^2}{C_0^2(z)} u(\mathbf{x}, \mathbf{s}, \omega). \tag{7.13a}$$

It should be noted that the equation for the scatter waves is nonlinear. When the scatter waves are much weaker compared with the incident wave, the above equation can be rewritten as follows after the Born approximation:

$$\left[\nabla^2 + \frac{\omega^2}{C_0^2(z)} \right] u_s(\mathbf{x}, \mathbf{s}, \omega) \approx -\alpha(\mathbf{x}) \frac{\omega^2}{C_0^2(z)} u_0(\mathbf{x}, \mathbf{s}, \omega), \quad \text{if } u_s \ll u_0 \tag{7.32}$$

Equation (7.12a) for the incident wave is completely the same as the equation satisfied by the Green function, and the expression of its solution has been defined in Section 5.4 after introducing coordinate transformation. The new coordinate variable τ is called the pseudotime and has three components as

$$\tau_1 = x/c(z), \quad \tau_2 = y/c(z), \quad \tau_3 = \frac{1}{b} \ln\left(\frac{a + bz}{a} \right) = \frac{1}{b} \ln\left(\frac{C_0}{a} \right)$$

The vertically variable velocity becomes

$$C_0(\tau_3) = ae^{b\tau_3}$$

In the new coordinates, the velocity C_0 varies exponentially with τ_3, and the corresponding Green function is proportional to waves u_0 but with a weight of $1/a$. Denoting τ' as the pseudocoordinates of the source, we get

$$G(\tau, \tau', \omega) = au_0(\tau, \tau', \omega) = \frac{e^{b\tau_3/2}}{4\pi|\tau - \tau'|} e^{\mp i\sqrt{\omega^2 - \left(\frac{b}{2}\right)^2}|\tau-\tau'|} \tag{7.33}$$

In a special case of a small vertical gradient of velocity, that is

$$\frac{b}{2} \ll \omega, \quad \text{i.e.} \quad \sqrt{\omega^2 - \left(\frac{b}{2}\right)^2} \approx |\omega|$$

the high-frequency approximation of the Green function becomes

$$G(\tau, \tau', \omega) = \frac{e^{b\tau_3/2}}{4\pi|\tau - \tau'|} e^{i\omega|\tau-\tau'|}$$

This expression indicates that the phase shift of high-frequency waves propagating in linearly variable velocity media in the pseudocoordinates is essentially identical to that propagating in homogeneous media in common coordinates. However, the geometric diffusion on wave amplitude shows differences. The vertical velocity gradient appearing should be considered as a weight to the geometric diffusion.

The Born approximation solution of Eqn (7.32) for the scattered waves can be expressed with the Green function. As shown in Eqn (5.19), the scattered waves

$$u_s(x, s, \omega) = \int_V G(x, x')F(s, x')dx'$$

$$= -\int G(x, x', \omega)\frac{\omega^2}{c_0^2(z)}\alpha(x')u_0(s, x', \omega)dx' \tag{7.34}$$

After the coordinate is transformed to the pseudotime system, the scattered waves becomes

$$u_s(s, \tau', \omega) = -\iint d\tau_1 \, d\tau_2 \int_{\tau_3 > 0} G(\tau, \tau', \omega)\frac{\omega^2}{c_0^2(\tau_3)}\alpha(\tau)u_0(s, \tau', \omega)d\tau_3 \tag{7.35}$$

The velocity C_0 in Eqn (7.35) is defined by Formula (5.35). By substituting the Green function Eqn (7.33) into Eqn (7.35), one obtains the

Born approximation solution with the scatter waves. For example, the approximation solution for high-frequency waves is

$$u_s(s, \tau', \omega) = -\iint d\tau_1 \, d\tau_2 \int\limits_{\tau_3 > 0} \frac{\omega^2 e^{b\tau_3/2}}{4\pi c_0^2(\tau_3)|\tau - \tau'|} e^{i\omega|\tau - \tau'|} \alpha(\tau) u_0(s, \tau', \omega) d\tau_3$$

(7.36)

where the incident wave field equals

$$u_0(\tau, s, \omega) = \frac{e^{b\tau_3/2}}{4\pi a|\tau - s|} e^{\mp i\sqrt{\omega^2 - \left(\frac{b}{2}\right)^2}|\tau - s|}$$

(7.37)

and the exponential sign depends on τ.

The linear Eqn (7.36) gives the Born approximation solution in linearly variant velocity media, and enables one to compute the square slowness disturbance α in seismic inversions. Giving a data set of scattered waves u_s and performing discretization of the slowness disturbance and the integral Eqn (7.36), we can apply the optimization criterion Eqn (7.3) to compute a generalized solution for inverse problems. The analytical solution of the inverse problem can be explicitly written after applying expanded Fourier transformation (Huang Lianjie and Yang Wencai, 1991).

To get the generalized solution with the equations shown in this section, the transformation from (x,y,z) to τ should be performed first. Because $\tau_3 = 0$ when $z = 0$, surface reflection waves u_s do not need to be transformed. Of course, the final estimated solution $\alpha(\tau)$ should be transformed back to $\alpha(x,y,z)$ after performing the inversion. As discussed in Section 6.5, whether the algorithm succeeds or not depends on the properties of the integral kernel at the right side of Eqn (7.36). The right side integral belongs to the oscillatory integral, and its convergence is conditional.

7.4. ACOUSTIC INVERSE SCATTERING PROBLEMS IN VARIANT VELOCITY MEDIA

We are going to discuss a special method for inverse scattering problems by using the inverse generalized Radon transform. However, we should study the forward generalized Radon transform first.

7.4.1. Acoustic Generalized Radon Transformation

The integral kernel of Born approximation solution Eqn (7.24) is the product of two functions: the incident wave $u_0(s, x', \omega)$ from a single point source and the Green function $G(x', x, \omega)$ that equals the incident wave of an imaginary source located at a receiver point. When the reference velocity $C_0(x)$ is a function of slowly variable and continuous, the incident waves $u_0(s, x', \omega)$ and Green function $G(x', x, \omega)$ can be determined using the WKBJ approximation mentioned in Section 5.5. The resulting approaches for inversion of the disturbance α (x) in a variable velocity background are called the Born–WKBJ approximation of inverse scattering (Sabatier, 1985; Yang Wencai, 1990, 1992, 1997).

The Born approximation solution (7.32) can be written as follows by the mean of the Green function

$$u_s(s, r, \omega) = \omega^2 \iint dx\, dy \int\limits_{z>0} \frac{G(r, x, \omega)G(s, x, \omega)}{c_0^2(x)} \alpha(x)dz \qquad (7.38)$$

where r represents the coordinates of a receiver point, $G(r, x, \omega)$ is the incident wave of the imaginary source at the receiver point, and $G(s, x, \omega)$ is the incident wave of the real single source. In WKBJ approximation, these waves can be expressed with an amplitude function A and phase function Φ as

$$G(x, y, \omega) = A(x, y)e^{i\omega\Phi(x,y)} \qquad (7.39)$$

The phase function Φ equals the traveling time between two points, satisfying the eikonal Eqn (5.56)

$$\nabla\Phi \cdot \nabla\Phi = C_0^{-2}(x)$$

while the amplitude function A satisfies the transport Eqn (5.57)

$$2\nabla\Phi \cdot \nabla A + A\nabla^2\Phi = 0$$

Substituting the eikonal equation and transport equation into Eqns (7.38) and (7.39), and introducing new parameters as

$$a(x, s, r) = A(x, s)A(x, r)/C_0^{-2}(x)$$

$$\Phi(x, s, r) = \Phi(x, s) + \Phi(x, r)$$

we have

$$u_s(s, r, \omega) = \omega^2 \iint dx\, dy \int\limits_{z>0} a(x, s, r)e^{i\omega\Phi(x,s,r)}\alpha(x)dz \qquad (7.40)$$

After inverse Fourier transformation to the time domain, we express scatter waves with a special integral form as

$$u_s(s, r, t) = \frac{\partial^2}{\partial t^2} \iiint a(x, s, r)\alpha(x)\delta[t - \Phi(x, s, r)]dx$$

$$= -\iiint a(x, s, r)\alpha(x)\delta''[t - \Phi(x, s, r)]dx \qquad (7.41)$$

For inversion of the square slowness disturbance $\alpha(x)$ in media of a variable background velocity, Eqn (7.41) gives a linear integral via the Born approximation. It can be used for seismic inversion of prestack common shot gathers. However, Eqns (7.40) and (7.41) belong to the Fredholm integral equation of the first kind with the oscillatory integral kernel, which has been discussed in Section 6.5.1 and causes instability in searching equation solutions.

As a matter of fact, Eqn (7.41) belongs to a variation of the generalized Radon transformation, called the acoustic generalized Radon transformation. The direct computation of the square slowness disturbance $\alpha(x)$ can be achieved by applying the inverse acoustic generalized Radon transformation. Assume that the amplitude function A varies slowly and that only two orders in the Taylor series expansion of the phase function Φ are taken, we can write (see the next section)

$$\alpha(x_0) = \frac{1}{8\pi^2 C_{\mathrm{out}}^3(x_0)}\left[1 + \frac{C_{\mathrm{out}}(x_0)}{C_{\mathrm{in}}(x_0)}\right]^3 \int\!\!\int \frac{|\cos \gamma(x_0, s, r)|^3}{a(x_0, s, r)} u_s(s, r, t = \Phi_0)\mathrm{d}n$$

$$(7.42)$$

where x_0 is an underground cell, C_{in} is the velocity in the neighborhood of the cell by the incident direction from source s, C_{out} is the velocity in the neighborhood of the cell by the side to receiver r, and γ is the half angle between the incident ray and the emergent ray. The integrals in Eqn (7.42) are performed along the surface $t = \Phi_0$ with its normal direction denoted by n. So far, we obtain the solution of the inverse scattering problem by applying the Born–WKBJ approximation method.

7.4.2. The Inverse Acoustic Generalized Radon Transformation

Now we are going to prove Eqn (7.42) and demonstrate inverse acoustic generalized Radon transformation. Based on the definition of the generalized Radon transformation , its inverse is the weighted integral of acoustic scattered waves along the surface $t = \Phi(x, s, r)$, that is

$$\alpha(x_0) = \int \mathrm{d}W(x_0, s, r)u_{s1}(s, r, t = \Phi(x_0, s, r))$$

$$(7.43)$$

where $\mathrm{d}W$ is an unknown function to be deduced, and x_0 is the center of an underground cell. The notation u_{s1} represents the pure reflection waves. In the theory of X-ray tomography, the Radon transformation is similar but with the integral performing along a plane (or a line in the 2D case)

$$u_s(\xi, p) = \int \alpha(x)\delta(p - \xi \cdot x)\mathrm{d}^3x$$

$$(7.44)$$

where u_s can be any scatter waves and is called the projection function in X-ray tomography, ξ is a vertical plane, and p is the normal distance from the origin to this plane, $p = \xi \cdot x$. The inverse Radon transformation can be expressed as

$$\alpha(x_0) = -\frac{1}{8\pi^2} \int d^2\xi \left[\frac{\partial^2}{\partial p^2} u_s(\xi, p) \right]_{p=\xi \cdot x_0} \tag{7.45}$$

Putting Eqn (7.44) into Eqn (7.45) yields

$$\alpha(x_0) = -\frac{1}{8\pi^2} \int d^2\xi \left[\frac{\partial^2}{\partial p^2} \int \alpha(x)\delta(p - \xi \cdot x) d^3x \right]_{p=\xi \cdot x_0} \tag{7.46}$$

$$= -\frac{1}{8\pi^2} \int d^2\xi \int d^3x \delta''(\xi \cdot (x_0 - x))\alpha(x)$$

where x is any point and x_0 is an image cell for tomography.

To perform inversion of velocity disturbance, the integral of the corresponding Radon transform should be performed along a curved surface (or curves in the 2D case), because the ray path is no longer straight. We shift the origin of the wave field to cell x_0, that is let $x = x_0 + y$, and then Eqn (7.44) can be rewritten as

$$u_{s1}(s, r, t) = -\int a(x_0 + y, s, r)\delta''[t - \Phi(x_0 + y, s, r)]\alpha(x_0 + y) d^3y \tag{7.47}$$

where the amplitude function a and phase function Φ are shown in Eqns (7.38) and (7.39), respectively. The Taylor expansion of Φ is

$$\Phi(x_0 + y, s, r) = \Phi(x_0, s, r) + [\nabla_x \Phi(x, s, r)]_{x=x_0} \cdot y$$

$$= \tau_0 + [\nabla_x \tau(x_0, s) + \nabla_x \tau(x_0, r)] \cdot y \tag{7.48}$$

It is known from the eikonal equation that

$$[\nabla_x \tau(x_0, s)]^2 = C_{in}^{-2}(x_0)$$

where subscript "in" is the velocity at a grid cell underground by the side of the incident ray (Figure 7.4), while

$$[\nabla_x \tau(x_0, r)]^2 = C_{out}^{-2}(x_0)$$

where subscript "out" denotes the velocity at the cell by the side of the departing ray. Figure 7.4 also shows a half angle γ between the incident and departing rays. Putting these equations into Formula (7.48) yields

$$\nabla_x \Phi(x_0, s, r) = 2 \left[\frac{1}{C_{in}(x_0)} + \frac{1}{C_{out}(x_0)} \right] \cos \gamma \xi(x_0, s \cdot r) \tag{7.49}$$

where ξ is the unit normal vector of the wave-front surface.

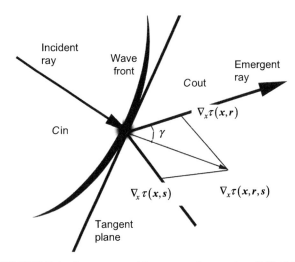

FIGURE 7.4   Description of Parameters in equation (7.45–7.55).

Assume that the amplitude function a at the neighboring cells is constant; substituting Eqn (7.49) into Eqn (7.47) yields

$$u_{s1}(s, r, t) = -a(x_0, s, r) \int \alpha(x_0 + y)\delta''[t - \tau_0 - k\xi(x_0, s, r) \cdot y]\mathrm{d}^3 y \quad (7.50)$$

where

$$k = -\frac{\cos \gamma}{C_{out}(x_0)}\left[1 + \frac{C_{out}(x_0)}{C_{in}(x_0)}\right] \quad (7.51)$$

Note that the integral in Eqn (7.50) is performed along the wave-front surface, whose tangent plane is denoted by $t - \tau_0 = k\xi(x_0, s, r) \cdot y$. Because $t = \tau_0 = \Phi(x_0, s, r)$ for x_0 and

$$\delta''(-bx) = \delta''(x)/|b|^3$$

Equation (7.50) becomes

$$u_{s1}(s, r, t = \tau_0) \approx -\frac{a(x_0, s, r)}{|k|^3} \int \alpha(x_0 + y)\delta''[\xi(x_0, s, r) \cdot y]\mathrm{d}^3 y \quad (7.52)$$

Now, we must remove the origin back from the imaging cell x_0 to any underground point x by $x = x_0 + y$. Thus, Eqn (7.52) can be rewritten as

$$u_{s1}(s, r, t = \tau_0) \approx -\frac{a(x_0, s, r)}{|k|^3} \int \alpha(x)\delta''[\xi(x_0, s, r) \cdot (x - x_0)]\mathrm{d}^3 x \quad (7.53)$$

Putting Eqn (7.53) into Eqn (7.43) yields

$$\alpha(x_0) \approx -\frac{a(x_0, s, r)}{|k|^3} \int dW(x_0, s, r) \int \alpha(x)\delta''[\xi(x_0, s, r)\cdot(x - x_0)]d^3x \quad (7.54)$$

By comparing Eqn (7.46) with Eqn (7.54), we find that the weight function in Eqn (7.43) should be

$$dW(x_0, s, r) = \frac{|k|^3}{8\pi^2 a(x_0, s, r)}d^2\xi \quad (7.55)$$

Finally, Eqn (7.42) can be obtained by putting Eqns (7.50) and (7.55) back into Eqn (7.43)

$$\alpha(x_0) \approx \frac{1}{8\pi^2 C_{out}^3(x_0)}\left[1 + \frac{C_{out}(x_0)}{C_{in}(x_0)}\right]^3 \int \frac{|\cos \gamma(x_0, s, r)|^3}{a(x_0, s, r)} u_{s1}(s, r, t = \tau_0)d^2\xi$$

It is the expression of the inverse acoustic generalized Radon transformation.

In practice, the inverse scattering inversion of using the acoustic generalized Radon transformation can be done after the discretization of the integral and gridding the cells with

$$\alpha(x_0) = \sum_k b(s, r, x_0)u(s, r, t = \Phi_k) \quad (7.42a)$$

Figure 7.5 shows the flow chart of a time-domain direct inversion procedure by input prestack CSP gather data. The inversion procedure involves 14 steps as follows:

1. Input prestack CSP gather data $u(s,r,t)$ and background velocity $C_0(x)$. $C_0(x)$ can be obtained by smoothing the interval velocity model.
2. Define the range to the square slowness disturbance α and the cell sequence j corresponding to discretized α_j.
3. Set $j = 1,2,\ldots,K$, for imaging cell cycles.
4. Initialize an accumulator by setting the sum $= 0$ and counter $L = 0$.
5. Repeat the computing loop of CSP, that is $s = 1,2,\ldots,m$.
6. For each receiver r, set $i = 1,2,\ldots,n$.
7. Compute the time shift $\Phi(s,r,x_0)$ and amplitude attenuation coefficient $b(s,r,x_0)$ at a given shot, receiver, and imaging cell by ray tracing.
8. Set $t' = \Phi + \Delta t$, $t' = \Phi + 2\Delta t,\ldots$
9. Sum up with Eqn (7.42a), sum $=$ sum $+ b(s,r,x_0)u(s,r,t')$, $L = L + 1$;
10. Go back to step 8 in cycles.
11. Go back to step 6 in cycles.
12. Compute the Born−WKJB solution estimates $\alpha_{jk} =$ sum$/L$.
13. Go back to step 3 and compute the square slowness disturbance $\{\alpha_j\}$ of all imaging cells in the range.
14. Convert $\{\alpha_j\}$ to reflectivity by using Eqn (7.10).

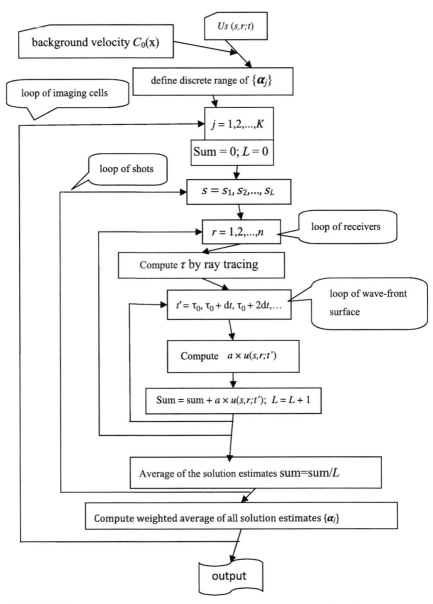

FIGURE 7.5   Flow chart of the time-domain direct inversion procedure by using inverse generalized Randon transform.

7.4.3. Some Supplements about Inverse Scattering Procedures

The last section demonstrates how to deduce an analytic solution of the inverse scattering problems by applying the Born–WKJB method. As a massive amount of data are involved in reflective seismic acquisition, inversion procedures should be accurate, stable, and of a high computational speed. The procedures based on analytic solutions, as shown above, usually have the highest computational speed. However, the Born−WKJB method originates from the perturbation theory, where $\alpha(x)$ should be much less than unity, meaning that there is only a weak velocity disturbance in $C(x,y,z)$ compared with the background velocity C_0. Actually, the reflection coefficient of the discontinuities in a sedimentary basin can reach ≥ 0.3. Whether the perturbation theory could be properly employed for the inversion of reflectors is under question (Tikhonov and Arsenin, 1977; Backus and Gilbert, 1970; Kaplan and Glass, 1995). In Section 6.5, we have shown that a pseudodifferential operator $\wp$ can be decomposed as the sum of an identical operator and a set of operators with different smoothness. If one wants only to locate discontinuities, which is equal to the first-order singularity, then one may take only the first one or two items in the expansion of operator $\wp$. The procedure of the inverse acoustic generalized Radon transform can be treated as a practical example of this concept. More accurate inversion methods for mapping higher-order singularities can be considered by using the idea of decomposition of the pseudodifferential operator.

The scattering theory can be extended to n-dimensional space. In the case of space R^n, assume that monopole sources and receivers are not overlapped. Denote Ω as the region containing scatters or reflectors in space R^n, and let its boundary be $\delta\Omega$. The location of monopole sources at the boundary can be marked as $x_s \in \partial\Omega$, that of the receivers is $x_r \in \partial\Omega$ and $x \in \Omega$ is any point in the area studied. Based on Eqn (7.13), the scatter waves satisfy

$$\nabla^2 u_s(x, x_s, \omega) + \left(\frac{\omega}{C_0}\right)^{n-1} u_s(x, x_s, \omega) = -\alpha(x)\left(\frac{\omega}{C_0}\right)^{n-1} \times [u_s(x, x_s, \omega) + u_0(x, x_s, \omega)] \quad (7.56)$$

If one uses the Green function in free spaces, the solution of u_s can be expressed by an integral, whose kernel is the product of the right items of Eqn (7.43) and the Green function. Considering the scatter wave generated from x to receiver x_r, the Hankel function for the acoustic waves can be extended into the space R^n, and the Green function can be written as

$$G_i(x, x_r) = -\frac{i}{4}\left(\frac{k}{2\pi|x - x_s|}\right)^{(n-2)/2} H^{(1)}_{(n-2)/2}(k|x - x_r|) \quad (7.57)$$

where $k = \omega/C_0$. Taking the first item in the asymptotic expansion of Hankel function of the first kind, we have

$$G_i(x, x_r) \approx e^{-i(\pi/2)(n+1)/2} \frac{k^{(n-3)/2}}{2(2\pi|x - x_r|)^{(n-1)/2}} e^{ik|x - x_r|} \quad (7.58)$$

This formula is accurate for the case of constant velocity. The incident wave $u_0(x, x_s)$ on the right side of Eqn (7.56) equals the wave field caused by a monopole source located at x. If one replaces x_s with x_r, then u_0 can be expressed by Eqn (7.58). In the case of weak scattering that $\alpha(x)$ varies slightly, the product $\alpha(x)u_s(x, x_s, \omega)$ on the right side of Eqn (7.56) can be omitted according to the Born approximation. Further assuming that C_0 is a constant, $k = \omega/C_0$, we have

$$u_s(k, x_r, x_s) \approx \frac{(-ik)^{n-1}}{4(2\pi)^{n-1}} \int_\Omega \frac{e^{ik|x-x_r|}e^{ik|x-x_s|}}{(|x-x_r||x-x_s|)^{(n-1)/2}} \alpha(x)dx \qquad (7.59)$$

where $\alpha(x)$ is an unknown for inversion. Equation (7.59) is an extension of the scattering theory in an n-dimensional space. If we set $n = 3$, then Eqn (7.59) becomes Eqn (7.36).

7.5. LINEARIZED ITERATIVE INVERSION OF SEISMIC REFLECTION DATA

The elegance of mathematical physics is accompanied by excessive simplification. The inverse scattering method of the Born approximation is not proper for propagation media with a large disturbance. More practical models are relevant to the operators for matching a large disturbance (Cohen and Bleistein, 1986; Tarantola, 1987; Yang Wencai, 1989; Beylkin and Rurridge, 1990; Yang Wencai and Du Jian-yuan, 1992, 1993), and more accurate solution estimates of inverse scattering can be successively approached by employing linear iterations. For the given two square slowness disturbance models α_1 and α_2, from Eqn (7.16), the difference between their reflection waves in the frequency domain is

$$\Delta U = u_2(s, r, \alpha_2) - u_1(s, r, \alpha_1)$$

$$= \iiint dy[\alpha_2(y)u_2(s, y) - \alpha_1(y)u_1(s, y)]G(r, y) \qquad (7.60)$$

where the Green function $G(y,r,\omega)$ equals the incident waves by an imaginary single source located at receiver r, and y is any point within the integration domain. If the iterative increment of the earth model $\Delta\alpha$ is very small, one can write

$$\alpha_1 = \alpha; \quad \alpha_2 = \alpha + \Delta\alpha \qquad (7.61)$$

The functional gradient of data with respect to the earth models can be defined using the Gateaux weak differential functional as

$$du(x, \alpha, f) = \lim_{\lambda \to 0} \frac{u(x; \alpha + \lambda f) - u(x; \alpha)}{\lambda} \qquad (7.62)$$

where f is the earth model. Substituting Eqn (7.61) into Eqn (7.60) and neglecting the high-order deviations yield

$$\Delta U = \iiint f(y)u_r(r, y, \alpha_1)u_s(s, y, \alpha_1)dy \qquad (7.63)$$

Equation (7.63) indicates linear dependency between the fitness, between the earth model and the input data, and the iterative modification increment of the earth model. Thus, an iterative procedure can be built based on the integral Eqn (7.63) to successively approach a generalized solution of inverse scattering problems. This procedure is called the linearized iterative inversion of seismic reflection data.

Seismic inversion computes square slowness disturbance $\alpha(x)$ by given reflection seismic data $u_{si}(i = 1, \ldots, I)$, and I the total number of reflection data gathers. The solution $\alpha(x)$ belongs to a generalized function. As the wave propagation operator contains a variable coefficient and is nonlinear, estimations of the solution can be decomposed into a series of linear approximation for iterative computation. Within each iteration, the nonlinear operator should be simplified into a linear one as expressed in Eqn (7.63) for practical computing.

Assuming that $\delta\alpha(x)$ is the variance of the earth model $\alpha(x)$, based on the definition of the functional Fréchet derivative, we have

$$B_i(\alpha + \delta\alpha) - B_i\alpha = \int F_i(y)\,\delta\alpha(y)dy \qquad (7.64)$$

where F_i is the Fréchet derivative of operator B_i with respect to $\alpha(x)$, Eqn (7.64) is linear for the unknown increment $\delta\alpha$. Corresponding reflection data increment is

$$\delta u_{si} = B_i(\alpha + \delta\alpha) - B_i\alpha \qquad (7.65)$$

It gives the wave field difference between two similar scatter models. We take Eqns (7.64) and (7.65) as restrictions and minimizing $\|\delta\alpha\|$, find the objective functional:

$$\phi = (\delta\alpha, \delta\alpha) + \sum_{i=1}^{I} \lambda_i[(F_i, \delta\alpha) - \delta u_{si}] \qquad (7.66)$$

where the brackets (,) represent the inner product as shown in Eqn (7.64), and λ_i is the Lagrange multiplier as shown in Eqn (7.64). Minimization of Eqn (7.66) yields

$$\delta\alpha = \sum_{i=1}^{I} \lambda_i F_i \qquad (7.67)$$

By considering Eqn (7.64), we get

$$\delta u_{si} = \sum_{k=1}^{I} (F_i, F_k)\lambda_i, \quad i = 1, \ldots, I \qquad (7.68)$$

where δu represents the difference between input seismic data and synthetic data calculated with iterative earth models. Multiplier $\{\lambda_i\}$ can be computed by solving the linear equation of Eqn (7.68), and the iteration increment $\delta\alpha(x)$ can be obtained by using Eqn (7.67).

Based on the above analysis, we have the linearized equation for the iteration as

$$\alpha_{q+1} = \alpha_q + \mu_q \sum_{i=1}^{I} \lambda_i F_{i,q}, \quad q = 0, 1, 2, \ldots \tag{7.69}$$

where q is the index of iterations, μ_q is the relaxation factor ranging from 0 to 1 at the qth iteration, and the initial mode α_0 should be a smooth function. The Fréchet derivative operator F_i can computed by using the method to be discussed in the next section, and the coefficient set $\{\lambda_i\}$ can computed by Eqn (7.67). As a result, the earth model $\alpha_q(x)$ can be modified iteratively according to Eqn (7.56), and finally be approached after the minimization of $\|\delta\alpha\|$.

To perform the linearized iteration inversion discussed above, the earth model $\alpha(x)$ must be discretized as a set of vectors:

$$\boldsymbol{\alpha}^T = \left(\boldsymbol{V}_1, \ldots, \boldsymbol{V}_j, \ldots\right) \tag{7.70}$$

Then, reflection seismic data can be discretized as

$$\boldsymbol{u}_{si} = \boldsymbol{B}_i\left(\boldsymbol{V}_1, \ldots, \boldsymbol{V}_j, \ldots\right)^T \tag{7.71}$$

where $j=1,2,\ldots,J$, J is the dimension of model parameters, the discretized earth model $\boldsymbol{V} \in \boldsymbol{R}^J$, limited in a finite dimensional Euclidean space. The corresponding Fréchet derivative is degenerated as a Jacobian matrix

$$F_{ij} = \left(\frac{\partial B_i}{\partial V_j}\right) \tag{7.72}$$

Using Eqn (7.64) yields

$$\delta \boldsymbol{u}_s = \boldsymbol{F}\delta\boldsymbol{V} \tag{7.73}$$

which is a linear equation set, and its damping least-square solution is

$$\delta\boldsymbol{V} = \left(\boldsymbol{F}^T\boldsymbol{F} + \lambda\boldsymbol{I}\right)^{-1}\boldsymbol{F}^T\delta\boldsymbol{u} \tag{7.74}$$

During the iteration with Eqns (7.69) and (7.74) can be rewritten as the iteration formula for the linearized inversion as

$$\boldsymbol{V}_{q+1} = \boldsymbol{V}_q + \mu_q\left(\boldsymbol{F}_q^T\boldsymbol{F}_q + \lambda_q\boldsymbol{I}\right)^{-1}\boldsymbol{F}_q^T\left(\boldsymbol{u}_s - \boldsymbol{u}_{s,q}\right) \tag{7.75}$$

where $q = 0, 1, 2\ldots$ is the index of iteration, $0 < \mu_q < 1$ is the regularization factor at the qth iteration, that is the damping factor or relaxation factor, and F_q is the Jacobian matrix at the qth iteration, λ_q is the Lagrange multiplier at the qth iteration. In addition, I is the unit matrix, u_s is the vector of input seismic data, and $u_{s,q}$ is the synthetic seismic data computed from the temporal earth model V_{q-1}.

Now, we look at the Fréchet functional derivative in Eqn (7.64) and find the relationship between changes of the model parameters and fitness of the data. For convenience, two similar models are denoted as $V_1(x)$ and $V_2(x)$ instead of as $\alpha_1(x)$ and $\alpha_2(x)$. The wave field difference of V_1 and V_2 can be computed by Eqn (7.63) as

$$\Delta u_s = u_s(x_r, x_s, t; \ V_1) - u_s(x_r, x_s, t; \ V_2)$$

$$= \int dt' \int dy[V_1(y)u_1(y, t') - V_2(y)u_2(y, t')]G(x_r, t|y, t') \qquad (7.76)$$

where u_1 and u_2 are the wave fields related to models V_1 and V_2 and excited by a single shot located at x_s, respectively. G equals the scatter waves coming from the scatter cell located at y and recording at point x_r. The background wave velocity C_0 is assumed for the wave propagation.

We denote a new function that is equivalent to an imaginary wave field of the scatter wave with a point source located at x_r:

$$\hat{u}(x, x_r, t; \ V) = G(x_r, t|x, t') + \int dt' \int dy \, \breve{G}(x, t|x, t')V(y)\hat{u}(y, x_r, t, V) \qquad (7.77)$$

where $\breve{G}$ is the conjugate Green function of G, that is the returning of waves,

$$\breve{G}(x, t|y, t') = G(x, -t|y, -t') \qquad (7.78)$$

For a clearer expression, hereafter time variable t will be neglected by writing $y = (y_1, y_2, y_3, t)$ as $y = (y_1, y_2, y_3)$, and dy contains the derivation with respect to time t. Note that u and $\hat{u}$ are both wave fields that can be computed by using the same synthetic procedures. Introduction of $\hat{u}$ helps us to calculate the functional gradient.

Assume that $\hat{u}_1$ is the imaginary wave field propagating in media $V_1(x)$. Putting this into Eqn (7.67) and arranging terms, we can use Eqn (7.76) to get

$$\Delta u_s = \int dy V_1(y)u_1(y)G(x_r|y) - \int dy V_2(y)u_2(y)\hat{u}_1(y)$$

$$- \int dy V_2(y)u_2(y) \int dy' \breve{G}(y|y')V_1(y')\hat{u}_1(y') \qquad (7.79)$$

where y and y' are two scatter cells in the media. Exchanging the integral order in Eqn (7.79) and exchanging y and y', Eqn (7.78) becomes

$$\Delta u_s = \int dy V_1(y) u_1(y) G(x_r|y) - \int dy V_2(y) u_2(y) \widehat{u}_1(y)$$
$$- \int dy V_1(y) \widehat{u}_1(y) \int dy' G(y|y') V_2(y') u_2(y') \tag{7.80}$$

Equation (7.60) shows that the last integration of the right side equals to $[u_2 - u_i]$, where u_2 is the total wave field (including incident waves and scattering waves) propagating in model V_2, and u_i represents the incident waves propagating in background media C_o and excited at x_s, with no relation to the earth model. Thus, we have

$$\Delta u_s = \int dy \Big[V_1(y) u_1(y) G(x_r|y) - V_2(y) u_2(y) \widehat{u}_1(y)$$
$$+ V_1(y) \widehat{u}_1(y) (u_2(y) - u_i) \Big] \tag{7.81}$$

Combining the second and third terms yields

$$\Delta u_s = \int dy [V_1(y) - V_2(y)] u_2(y) \widehat{u}_1(y)$$
$$+ \int dy \Big[V_1(y) u_1(y) G(x_r|y) - V_1(y) \widehat{u}_1(y) u_i(y) \Big] \tag{7.82}$$

The Green function equals the total wave field excited by a single impulse subtracting other waves

$$G(x_r|y) = \widehat{u}_1(y, V_1) - \int dy \ \breve{G}(x_r|y) V_1(y) \widehat{u}_1(y, V_1) \tag{7.83}$$

Similarly,

$$u_i(y) = u_1(y, V_1) - \int dy \ G(x_s|y) \ V_1(y) u(y, V_1) \tag{7.84}$$

Substituting Eqns (7.83) and (7.84) into Eqn (7.82) yields

$$\Delta u_s = \int dy [V_1(y) - V_2(y)] u_2(y) \widehat{u}_1(y)$$
$$+ \int dy \Big[V_1(y) \widehat{u}_1(y) u_{1s}(y) - V_1(y) u_1(y) \widehat{u}_{1s}(y) \Big] \tag{7.85}$$

where u_{1s} is the scatter wave excited by a source located at x_s, that is

$$u_{1s} = \int dy' \ G(x_s|y') V_1(y') u_1(y') \tag{7.86}$$

and

$$\hat{u}_{1s} = \int dy' \; \breve{G}(x_r|y')V_1(y')\hat{u}_1(y') \tag{7.87}$$

One might find that the second term on the right side of Eqn (7.85) can be negligible for deep reflective waves, because the distance between x_r and x_s is far smaller than the distance from them to the deep reflectors. As u_{1s} increases with u_1 and $\hat{u}_{1s}$ increases with $\hat{u}_1$, and the second term on the right of Eqn (7.85) can be canceled, leading to

$$u_s(x_r, x_s, V_1) - u_s(x_r, x_s, V_2) \cong \int dy[V_1(y) - V_2(y)]u_2(y)\hat{u}_1(y) \tag{7.88}$$

By denoting $V_1 = V_2 + \delta V$, the Fréchet derivative that is defined in Eqn (7.64) becomes

$$F(y, x_r, x_s, V) \cong \hat{u}_1(y, x_r, V)u_2(y, x_s, V) \tag{7.89}$$

By referring Eqn (7.75), the expression of Eqn (7.89) for the linearized iteration becomes

$$F_{i,q} \equiv F(y, x_r, x_s, V_q) = \hat{u}_1(y, x_r, V_q)u_2(y, x_s, V_q) \tag{7.90}$$

where V_q is the square slowness disturbance in the qth iteration, $\hat{u}_1$ is the imaginary wave field computed from V_q, and u_2 is the real wave field computed from V_q. So far, we can compute the functional derivative by using any forward synthesis methods plus the product operation as shown in Eqn (7.90). Going back to the linear iteration Formula (7.75)

$$V_{q+1} = V_q + \mu_q \left(F_q^T F_q + \lambda_q I\right)^{-1} F_q^T \left(u_s - u_{s,q}\right),$$

where the functional derivative can be computed by Eqn (7.90). Now we may turn to a simplified case of nonlinear iterations by modifying the model weight as

$$\delta V_q(x) = \left(F_q^T F_q + \lambda_q I\right)^{-1} F_q^T \left(u_s - u_{s,q}\right), \quad q = 1, 2... $$

where V_q is the square slowness disturbance in the qth iteration, δV_q is its modification increment, and μ_q is a relaxation factor. Then, Eqn (7.75) becomes

$$V_q(x) = V_{q-1}(x) + \mu_q \delta V_q(x), \quad q = 1, 2... \tag{7.91}$$

By iterative modification of the earth model with Eqn (7.91), one has to analyze the mean-square error of data fitness to evaluate an approximate solution estimate.

The computed reflection data in the qth iteration corresponding to Eqn (7.76) is

$$u_s(x_r, x_s; V_q) \cong u_s(x_r, x_s; V_{q-1}) + \mu_q \Delta U(x_r, x_s) \tag{7.92}$$

where x_r is the receiver point, x_s is the shot, and ΔU is the difference in reflection wave fields between two iterations. We should remember that Eqn (7.92) is an approximation in which high-order terms have been ignored. ΔU is expressed as

$$\Delta U(x_r, x_s) = \int dy \, \delta V_q(y) \widehat{u}(y, x_r; V_{q-1}) u(y, x_r; V_{q-1}) \tag{7.93}$$

and can be computed from the earth model V_{q-1}.

To find a generalized solution, we have to study the mean-square errors of the fitness. Assume that $V(x)$ is the real earth model that produces accurate reflection data $u_s(x_r, x_s; V)$. The fitting errors at the qth iteration equal

$$e(x_r, x_s; V_q) = u_s(x_r, x_s; V_q) - u_s(x_r, x_s; V) \tag{7.94}$$

Substituting Eqn (7.92) yields

$$\begin{aligned} e(x_r, x_s; V_q) &= u_s(x_r, x_s; V_{q-1}) + \mu_q \Delta U(x_r, x_s) - u_s(x_r, x_s; V) \\ &= e(x_r, x_s; V_{q-1}) + \mu_q \Delta U(x_r, x_s) \end{aligned} \tag{7.95}$$

Based on the definition, the mean-square fitting errors in the linearized inversion are

$$E(V_q) = \iiiint dx_s dx_r dx_r' dx_s' \, e^*(x_r, x_s; V_q) W(x_r, x_s; x_r', x_s'; V_q) e(x_r', x_s'; V_q) \tag{7.96}$$

where the star denotes the conjugate function, and W is the covariance matrix of the data. Assume that the data errors follow a Gaussian distribution and are independent of each other; then,

$$W(x_r, x_s; x_r', x_s') = W(x_r, x_s) \delta(x_r - x_r') \delta(x_s - x_s')$$

So Eqn (7.96) becomes

$$\begin{aligned} E(V_q) &= \iint dx_s dx_r W(x_r, x_s) |e(x_r, x_s; V_q)|^2 \\ &= \iint dx_s dx_r W(x_r, x_s) |e(x_r, x_s; V_{q-1}) + \mu_q \Delta U(x_r, x_s)|^2 \\ &= E(V_{q-1}) - 2\mu_q P_q + \mu_q^2 Q_q \end{aligned} \tag{7.97}$$

where

$$P_q = -\text{Re} \iint dx_s dx_r W(x_r, x_s) e(x_r, x_s; V_{q-1}) + \Delta^* U(x_r, x_s) \qquad (7.98)$$

$$Q_q = \iint dx_s dx_r W(x_r, x_s) |\Delta^* U(x_r, x_s)|^2 \qquad (7.99)$$

The mean-squares $E(V_q)$ can be computed by substituting Eqn (7.93) into Eqn (7.97).

To minimize the mean-square error, $E(V_q)$ in Eqn (7.97) can be differentiated to μ_q. If $dE/d\mu_q = 0$, then it yields

$$\mu_q = P_q / Q_q \qquad (7.100)$$

By putting μ_q back into Eqn (7.97), we have

$$E(V_q) = E(V_{q-1}) - P_q^2 / Q_q \qquad (7.101)$$

Obviously, the minimization procedure requires P_q to be maximal and Q_q to be minimal at the qth iteration.

To maximize P_q and to minimize Q_q for the earth model $V(x)$, one must find their partial derivative with respect to V. The Gateaux functional derivative can be used once more, and using Eqn (7.99) yields

$$
\begin{aligned}
dE(V) &= \lim_{\lambda \to 0} \frac{E(V + \lambda \delta V) - E(V)}{\lambda} \\
&= 2\text{Re} \iint dx_s dx_r W(x_r, x_s) e(x_r, x_s, V) e^*(x_r, x_s; \delta V) \\
&= 2 \int dx\, \delta V(x) g(x, V) \qquad (7.102)
\end{aligned}
$$

where

$$g(x, V) = \text{Re} \iint dx_s dx_r W(x_r, x_s) e(x_r, x_s; V) \hat{u}^*(x, x_r; V) u^*(x, x_s; V) \qquad (7.103)$$

Specially in the qth iteration,

$$g(x, V_q) = \text{Re} \iint dx_s dx_r W(x_r, x_s) e(x_r, x_s; V_q) \hat{u}^*(x, x_r; V_q) u^*(x, x_s; V_q) \qquad (7.104)$$

Substituting Eqns (7.84) and (7.104) into Eqn (7.98) yields

$$P_q = -\int dy\, \delta V_q(y) g(y, V_{q-1}) \qquad (7.105)$$

It can be known from Eqn (7.101) that maximum P_q leads to $E(V_q)$ becoming a minimum, and it requires that the two integral kernels in Eqn (7.105) turn to overlap as

$$\delta V_q(x) = -g_{q-1}(x) \equiv -g\left(x, V_{q-1}\right) \tag{7.106}$$

Equivalently,

$$P_q = \int dx g_{q-1}^2(x) \tag{7.107}$$

Accordingly, the iterative linearized inversion procedure should combine Eqn (7.100) for minimum Q_q and Eqn (7.106) for maximum P_q. Putting them into Eqn (7.91) yields

$$V_q(x) = V_{q-1}(x) - \frac{P_q}{Q_q} g_{q-1}(x), \tag{7.108}$$

which represents the steepest descent procedure for the nonlinear inversion.

In addition, we can make P_q maximum and Q_q minimum simultaneously, and yield another procedure for the iterative linearized inversion by minimizing the mean-square error. Similar to Eqn (7.108), the modification of the earth model in the qth iteration becomes

$$\delta V_q(x) = -g_{q-1}(x) + \frac{P_q}{P_{q-1}} \delta V_{q-1}(x), \quad q > 1 \tag{7.109}$$

and

$$V_q(x) = V_{q-1}(x) + \frac{P_q}{Q_q} \left[-g_{q-1}(x) + \frac{P_q}{P_{q-1}} \delta V_{q-1}(x) \right] \tag{7.110}$$

This procedure belongs to the conjugate gradient descent method in computational mathematics.

Based on the above discussion, a procedure for the linearized iterative inversion by minimizing the mean-square fitness can be briefly summarized as follows:

Step 1. Set the average background velocity and the initial model of the square slowness disturbance $V_0(x)$, and then compute the data fitting error $e_0(x_r, x_s, t)$, the wave field u_0 and $\hat{u}_0$. Compute $g_0(x)$ and the mean-square fitness error $E(V_0)$.

Step 2. In the iteration of $q = 1, 2, ...$, compute P_q, δV_q, ΔU, Q_q, V_q, e_q, g_q, and $E(V_q)$ in order. The computation has been discussed before.

Step 3. Go back to Step 2 to perform the $(q + 1)$th iteration, or stop the iteration according to some given criteria.

The linearized iterative inversion can misfit the global minimum sometimes, because the final estimate is more or less dependent on the initial model. New nonlinear inversion methods are needed for comparison, such as stochastic inversion methods.

7.6. THE MAXIMUM ENTROPY INVERSION AND INVERSION FOR RESERVOIR PARAMETERS

Seismic inversion is a kind of deductive reasoning that tries to understand real earth models based on physical equations and observed seismic data. The construction of a concept about the model relies on the basic information, such as the observed boundary conditions, the accuracy of constitutive equations, and kinematic equations. Denote I as the basic information and H as the event of constructing the concept H. According to the theory of probability, orderly deduction of H follows a special rule called Bayes' theorem, which involves computing the Bayes' prior probability and the posterior probability. Inversion methods based on Bayes' theorem will be discussed as follows.

7.6.1. Bayes' Theorem and Maximum Entropy Inversion

It is believed in the probability theory that the motion of all matter existing in nature has an uncertainty, so the probability of any natural event cannot be 100%. Thus, the probability of the solution estimates that approach the accurate solution is certainly <1. One may consider the maximization of the probability density for the construction of new inversion procedures. Proposed by Tarantola in 1984, the probability density optimization inversion method has attracted more attention gradually (Ulrych et al., 1992) and be applied to reservoir parameter analysis. Assume that y is a model parameter, $x(y)$ is an unknown model related to a certain physical experiment, and denote $\tilde{d}_i$ as a set of observation data, $i = 1,2,...,N$. As the data has an error δ whose covariance matrix is,

$$\mathbf{Covar}\left[\delta_i, \delta_j\right] = \left[C_{ij}\right].$$

The unknown model $x(y)$ can be discretized into a set of parameters $\{x_k, k = 1, 2, ..., M\}$. If the data has a linear relationship with the model, then it can be written as

$$\tilde{d}_i = \sum_k G_{ik} x\left(y_k\right) + \delta_i \tag{7.111}$$

where the $N \times M$ matrix G is a linear operator with elements as

$$G_{ik} = G_i(x_k)(x_{k+1} - x_{k-1})/2. \tag{7.112}$$

How does one evaluate the solution estimates of Eqn (3.111)? In the probability theory, the characteristics of solution estimates can be measured with χ^2 distribution as

$$\chi^2 = \sum_{i=1}^{N} \sum_{j=1}^{N} \left[\tilde{d}_i - \sum_{k=1}^{M} G_{ik}\tilde{x}(y_k) \right] C_{ij}^{-1} \left[\tilde{d}_j - \sum_{j=1}^{M} G_{jk}\tilde{x}(y_k) \right] \qquad (7.113)$$

If nondiagonal elements in the matrix are ignored, then,

$$\chi^2 \approx \sum_{i=1}^{N} \left[\frac{\tilde{d}_i - \sum_{k=1}^{M} G_{ik}\tilde{x}(y_k)}{\sigma_i} \right]^2 , \sigma_i \equiv (\mathbf{Covar}[i,i])^{1/2}. \qquad (7.114)$$

As a result, if solution estimates $\tilde{x}$ follows the maximum concentration theorem of the χ^2 distribution, it must be the solution estimate that has the least effect caused by data variance. So, a new criterion for the inversion can be expressed as

$$\chi^2[\tilde{x}] + \lambda(\tilde{x} \cdot \tilde{x}) = \min \qquad (7.115)$$

The generalized solutions obtained under this criterion are called zero-order regularization solutions.

Now, we may apply the Bayes' probability theorem to seismic inversions (Riley, 1974; Ulrych et al., 2001). The theorem indicates that for given two sets of related events A and B, after the occurrence of event B, the conditional probability (posteriori probability) of event A equals

$$\mathbf{Prob}(A|B) = \mathbf{Prob}(A)\frac{\mathbf{Prob}(B|A)}{\mathbf{Prob}(B)} \qquad (7.116)$$

where $\mathbf{Prob}(B|A)$ is the prior probability, that is the conditional probability of event B after a given event A.

The inversion solution estimates show a concept for the interpretation of the seismic data. An index to evaluate the accuracy of this concept is called plausibility. Bayes' theorem can be used to calculate the plausibility of a theory, a hypothesis, or a proposal. Construction of any concept about an earth model relies on some basic information, including the boundary and initial conditions, the constitutive equations, and kinematic equations. We denote the basic information as I, and H is the event to create the concept about the earth model, then $\mathbf{Prob}(H|I)$ is the conditional probability of building the right concept, called the Bayes prior probability. The more accurate the forward physical model accepted, the larger the Bayes' prior probability, and the higher the inversion plausibility will be. The inversion plausibility needs further verification by experiments. Combining experimental data D_a and Bayes' theorem, the plausibility of an event H becomes

$$\textbf{Prob}(H|D_a I) = \textbf{Prob}(H|I)\frac{\textbf{Prob}(D_a|HI)}{\textbf{Prob}(D_a|I)}. \tag{7.117}$$

The denominator $\textbf{Prob}(D_a|I)$ on the right side of Eqn (7.117) is the probability of the observed data of the experiment with the model, and has relation with experimental precision. The larger the error in experimental data, the lesser the possibility of using it to verify the inversed model, and the lower the inversion plausibility will be. The numerator $\textbf{Prob}(D_a|HI)$ on the right side of Eqn (7.117) is the prior prediction probability of data, which turns to a standard constant to make the summation of the probability of coming from different ideas equal to unity. In seismic inversion, $\textbf{Prob}(H|D_a I)$ is the probability of obtaining real earth models according to observed data and forward equations.

Applying Bayes' theorem to inversion of a set of elastic parameters, written as x, we treated seismic elastic parameters as the concept of the earth models. Denote the input data as d, wave propagation operator and the source in equation as I. Because the data are limited, the operator has been simplified, and the inverse problem can be ill posed, the solution estimate x is usually unique, that is it is not located at a point in model space, but in a region with a certain size. This region for all possible x is called a posterior bubble. Bayes' theorem (7.116) gives the criterion to fix a proper point x that has the highest plausibility within the posterior bubble. If the data error obeys Gaussian distribution, then we can get

$$\textbf{Prob}(d|xI) = \exp\left(-\frac{1}{2}\chi^2\right)\Delta x_1 \Delta x_2 \cdots \Delta x_m \tag{7.118}$$

where χ^2 can be computed with Eqn (7.113) or Eqn (7.114), and Δx is the interval of the elements of x.

If $\textbf{Prob}(x|I)$ in Eqn (7.117) equals a constant, then

$$\textbf{Prob}(x|dI) = \beta \frac{\textbf{Prob}(d|xI)}{\textbf{Prob}(d|I)}. \tag{7.119}$$

The maximum entropy method (MEM) uses Eqn (7.119) in its computation. If one looks at the specific procedure of the MEM method in detail, one has to see Shannon's theorem first, in which entropy E of a physical dynamical system is determined by the negative value of the system information I. The so-called negentropy $H = I - \max(E)$ is the difference between the system information and the maximum entropy. Bayes' prior probability $\textbf{Prob}(x|I)$ is proportional to $\beta = \exp(-H)$. The goal of MEM inversion is to maximize the negative entropy, that is the negentropy H.

The first step is discretizing the solution estimate of the unknown earth model as vector x that is located at in m-dimensional space and has components denoted with index k, $k=1,2,\ldots m$ convert it into integer form by classification. Its norm is defined as

$$X = \sum_{k=1}^{m} |x_k| \tag{7.120}$$

Using Eqn (7.120) and the prior probability, the negative entropy satisfies

$$H(\mathbf{x}) = \sum_{k=1}^{m} x_k \ln(x_k/X). \tag{7.121}$$

Putting Eqns (7.113) and (7.121) into Eqn (7.119) yields

$$\mathbf{Prob}(\mathbf{x}|\mathbf{d}) \propto \exp\left[-\frac{1}{2}\chi^2\right]\exp[-H(\mathbf{x})] \tag{7.122}$$

Maximization of Eqn (7.122) is equivalent to minimizing

$$-\ln[\mathbf{Prob}(\mathbf{x}|\mathbf{D})] = \frac{1}{2}\chi^2[\mathbf{x}] + H[\mathbf{x}] = \frac{1}{2}\chi^2[\mathbf{x}] + \sum_{k=1}^{m} x_k \ln(x_k/X) \tag{7.123}$$

By combining the constants and X into the regularization factor, criterion (7.123) can be rewritten:

$$\Phi_s = \chi^2[\tilde{\mathbf{x}}] + \lambda \sum_{k=1}^{m} \tilde{x}_k \ln \tilde{x}_k = \min \tag{7.124}$$

where the head script "$\sim$" represents the solution estimate. The criterion (7.124) belongs to the MEM. Both the negative entropy in Eqn (7.121) and the functional Φ in Eqn (7.124) have a nonlinear relationship with the estimated solution x as a logarithm function appeared. If the earth model x varies sharply, a better solution estimate can be achieved by employing the MEM.

7.6.2. Probability Density Inversion Based on Statistical Estimation of Rock Physical Properties

We have discussed inversion theory and methods in this chapter for computing elastic parameters of underground rocks. The achievements of reflection seismic exploration encourage the expansion of its application to the delineation of hydrocarbon reservoirs, and to enhance oil/gas production. After 1990's, the petroleum industry has been requiring computation of reservoir parameters, including porosity, permeability, and fluid parameters.

The thickness of oil/gas reservoirs is usually less than half the wavelength of reflected waves. The reservoirs are saturated with fluid; the elastic wave equation had been studied by Gassmann and Biot as shown in Chapter 3. Biot's theory shows that there is stress–strain coupling between a solid matrix and the pore fluid because of friction that occurs at their boundaries. As a result, the second type P wave is created, which attenuates very quickly and distributes in a small range. What must be the interaction between the P wave of the second type and the reservoir boundaries? Is it possible that more seismic waves with a microsize could be created by the interaction?

Besides pored reservoirs, scientists have also studied fracture-type reservoirs. In 1970s, Berryman has deduced the elastic wave equations that can be applied to fracture-type reservoirs. However, the characterization of reservoirs must be based on seismic inversions. The elastic parameters should be inverted first, then reservoir parameters would be computed by applying statistical results that link petrophysical properties and the elastic parameters (Bosch et al., 2010). Such a process is called the probabilistic estimation of petrophysical properties and seismic inversion.

Bayes' theorem (7.116) for computing accurate reservoir parameters can be rewritten as

$$\mathbf{Prob}(R, B) = \mathbf{Prob}(R)\frac{\mathbf{Prob}(B|R)}{\mathbf{Prob}(B)} \qquad (7.116a)$$

where a specified event B is the inversion of elastic parameters based on given seismic data, and event R is the generation of reservoir properties. Then, the prior probability $\mathbf{Prob}(B|R)$ is the probability of inverting elastic parameters under proper petrophysical property statistics. Posterior probability $\mathbf{Prob}(R, B)$ is the probability of inverting elastic parameters and reservoir parameters simultaneously based on seismic data when reservoirs are in existence.

The probabilistic estimation of petrophysical properties and seismic inversion has been developed based on Bayes' theorem. The basic idea will be explained as follows: Posterior probability density (PPD) can be determined by the likelihood function. The maximum PPD yields the solution estimate with the highest plausibility. Denote m as the reservoir parameters and $PPD(m)$ as the PPD of an inverting reservoir parameter m; from Bayes' theorem, the $PPD(m)$ is proportional to the prior probability density $P(m)$, that is the accuracy of petrophysical property estimations. On the other hand, the $PPD(m)$ is inversely proportional to the norm of the fitness between seismic data and forward synthetic wave field $|d - g(m)|$, that is proportional to the likelihood function A as

$$PPD(m) = P(m)A(d - g(m)) \qquad (7.125)$$

The prior probability density $P(m)$ in Eqn (7.125) can be computed based on Bayes' theorem as

$$P(m) = P(m_{elas}|m_{res})P(m_{res})$$ (7.126)

where m_{res} represents the reservoir parameters, including porosity, permeability, fluid factor, m_{elas} represents the elastic parameters, including elastic wave velocities, Lame constants, and $P(m_{res})$ is the probability of the emergence of the correct reservoir parameters, and $P(m_{elas}|m_{res})$ is the probability of the emergence of elastic parameters under given reservoir parameters. Putting Eqn (7.126) into Formula (7.125) yields

$$PPD(m) = P(m_{elas}|m_{res})P(m_{res})A(d - g(m))$$ (7.127)

Equation (7.127) states that a good result on reservoir characterization depends not only on the likelihood A of seismic data but also on the accuracy of petrophysical property estimation. Maximum $PPD(m)$ gives inversion estimates of reservoir parameters with the highest plausibility.

7.7. SUMMARY

To find generalized solution estimates for seismic inversion it requires criteria that can be expressed mathematically. Some different criteria have been discussed in this chapter, producing different results for the interpretation of seismic data. Which criterion is the best for application? This question is hard to answer because each of them has different properties and should be chosen according to the amount of data, observation errors, and the complexity of the earth models. Theoretically, the solution estimate of the largest plausibility belongs to a universal criterion, and can be accepted for seismic inversion to some extent.

Generalized linear inversion methods are mature in solving linear inverse problems (Menko, 1984; Yang Wencai, 1989, 1997). Nonlinear inversion problems can be solved using the linearized iterative inversion method that has a good mathematical foundation. Some problems exist in this type of inversion procedures, including the parameterization of complex earth models, improvement of forward modeling, acceleration of the calculation of the inverse operator, convergence and control of the iteration process, and multiple searching starting from the initial model, etc. Practical inversion procedures must have a high computation speed and stability.

In this chapter, we have discussed only the inversion of acoustic equations, which already involves the Fourier integral operators or pseudodifferential operators. More complicated earth models, such as anisotropic models or viscoelastic models, contain much more unknown parameters in wave equations, resulting in heavy uniqueness in seismic inverse problems. We need further innovations on the seismic inversion theory to build solid foundations for the inversion of elastic wave equations as required by reservoir delineation.

Appendix: Finite Difference Method for Solving the Acoustic Wave Equation with Velocity and Density Variant Media

A1. INTRODUCTION

In order to train graduate students on computational skills, this book encloses a source program for solving acoustic wave equation by using the finite difference method. We do not introduce details of the finite difference and finite element methods that have been discussed by many books (Boore, 1972; Riley, 1974; Beyer, 1981; Silvester, 1983; Nedoma, 1999). This Appendix only shows how to apply the finite difference method for reflection seismic modeling and how to use the program.

Many excellent programs of the finite difference method for reflection seismic modeling are available today, but they may not be convenient for quick users because construction of the Earth models consumes a lot of time. In the following programming, we employ autogridding techniques to reduce input data, reducing most of the time for the data input. Both the five-point and the nine-point finite difference schemes can be employed in the programming as controlled by an option factor. The nine-point finite difference scheme usually owns better accuracy in seismic modeling.

A2. MATHEMATICAL MODELS OF SEDIMENTARY BASIN

A sedimentary basin should be mathematically described by a few parameters and vectors for modeling. The parameters involved include the interface number, interface depth, interval velocity, and density of each layer, which are functions of variable x in two-dimensional cases. The layered model has been introduced in Section 4.7. The index of the interface matches the index of the layer above. The varying density can be calculated from varying velocity according to the Gardner formula (see Sheriff, 1984). Therefore, the input contains the depth matrix $H(i, j)$ and the velocity matrix $V(i, j)$ only. The depth matrix $H(i, j)$ can describe structures like pinchout, fault, and traps as follows.

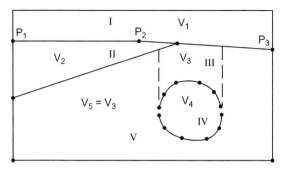

FIGURE A.1   A cylinder model with five interfaces. Interface I: defining points P_1–P_3.

For example, a cylinder model can be described with five interfaces as shown in Figure A.1. On an (x, z) plane, interfaces are functions $\{H_i = f_i(x),$ $0 \leq x \leq b\}$, $i = 1, 2\ldots$ We must define several defining points for each interface and then use interpolation to calculate discrete $\{x_i\}$. A cylinder must contain two interfaces for describing as shown in Figure A.1, using 12 defining points. The regulations for the discrete modes include that one must define an interface and its upper-side interval velocity simultaneously. Do not use vertical lines for the interfaces as shown in Figure A.2, which describes a fault in the profile. Finally, the bottom of the model must be treated as the last interface, although its reflection might not be recorded if the record length is not long enough. Figure A.3 shows how to describe a diapir fold model.

The input format of Earth models is stated as follows. The first two lines define the size of the profile. Unit of velocity is meters per second.

Line 1: horizontal start point (m), end point (m), scale of x
Line 2: vertical start point (m), end point (m), scale of z
Line 3: the total number of layers
Line 4: Layer 1, node point number, V_p (for constant velocity layer), V_s
Line 5: Layer 2, node point number, V_p (for constant velocity layer), V_s

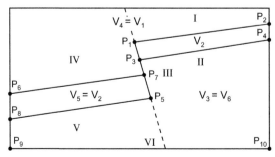

FIGURE A.2   A fault model. Interface I, P_1–P_2; interface II, P_1–P_3–P_4; interface III, P_3–P_7; interface IV, P_6–P_7; interface V, P_8–P_5; and interface VI, P_9–P_{10}.

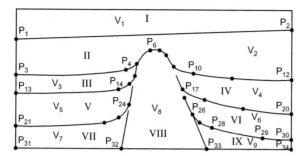

FIGURE A.3   A diapir fold model. Interface VII, $P_{31}-P_{32}-P_{24}$; interface VIII, $P_{32}-P_{33}$; and interface IX, $P_{26}-P_{33}-P_{34}$.

Line 6: Layer 3, node point number, V_p (for constant velocity layer), V_s
Line 7: Layer 4, node point number, V_p (for constant velocity layer), V_s
......

A3. SEISMIC MODELING WITH THE FINITE DIFFERENCE METHOD

The seismic modeling produces reflection records for common-midpoint (CMP), common shot point (CSP) gathers, stack profiles, vertical seismic profiling (VSP) records, and cross-hole seismic records under given Earth models.

A3.1. Seismic Modeling Formula

The equation used is the acoustic wave equation with variable density ρ and velocity C

$$\frac{1}{C^2}\frac{\partial^2 u}{\partial t^2} - \nabla\left(\frac{1}{\rho}\nabla u\right) = \delta(r,s)f(t) \tag{A.1}$$

where u is pressure wavefield. The density is calculated by Gardener formula as

$$\rho = 309.5\sqrt[4]{C} \tag{A.2}$$

where units are meters per second for velocity and kilogram per meter cube for density. Letting s be source locations and r be receivers, expansion of (1) yields

$$\frac{\partial^2 u}{\partial t^2} = C^2\left[\left(\frac{\partial^2 u}{\partial x^2}+\frac{\partial^2 u}{\partial z^2}\right)-\frac{1}{\rho}\left(\frac{\partial \rho}{\partial x}\frac{\partial u}{\partial x}+\frac{\partial \rho}{\partial z}\frac{\partial u}{\partial z}\right)\right]+\delta(r,s)f(t) \tag{A.3}$$

The boundary condition on the free surface for stress is

$$\sigma_{zz}^{(1)}|_{z=0} = 0 \tag{A.4}$$

On the interface between layers 1 and 2, the boundary conditions for stress and wavefield are

$$\sigma_{zz}^{(1)} = \sigma_{zz}^{(2)} \tag{A.5}$$

$$u_1 = u_2 \tag{A.6}$$

The initial conditions are

$$u|_{t\leq 0} = 0$$

$$\partial u/\partial t|_{t\leq 0} = 0 \tag{A.7}$$

Conditions (A6) and (A7) are automatically set during performing the finite difference procedure.

A3.2. The Finite Difference Procedure

Let the size of grid cells $\Delta X = \Delta Z = h$, horizontal index of a node of the grid be m, and vertical index be n, then a node $(x, z) = (mh, nh)$ can be dictretized as (m, n).

1. Second-order difference respect to time

$$\left.\frac{\partial^2 u}{\partial t^2}\right|_{(x_0,z_0,t_0)} \approx \frac{u_{t_0+\Delta t} - 2u_{t_0} + u_{t_0-\Delta t}}{\Delta t^2} \tag{A.8}$$

where t_0 is the excitation moment, Δt is the time interval, and wavefield $u_t = u(x_0, z_0, t_0)$.
2. Second-order difference with respect to space
The second-order difference can be divided into the five-point and nine-point schemes. We will discuss these as follows:
a. The five-point central difference scheme (see Figure A.4) uses

$$\nabla^2 u \approx \frac{a_1 u_{m,n} + a_3 (u_{m+1,n} + u_{m,n+1} + u_{m-1,n} + u_{m,n-1})}{h^2} \tag{A.9}$$

where the coefficients equals

$$\begin{bmatrix} & a_3 & \\ a_3 & a_1 & a_3 \\ & a_3 & \end{bmatrix} = \begin{bmatrix} & 1 & \\ 1 & -4 & 1 \\ & 1 & \end{bmatrix}$$

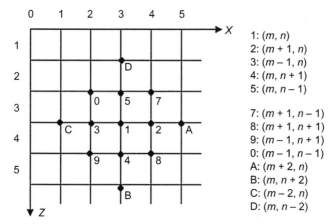

FIGURE A.4 The finite difference grid and node coding. Points 1–5 for five-point central difference, points 0–9 for nine-point second-order difference (block mode), and points 1–5 plus A–D for nine-point fourth-order difference (cross-mode).

b. The nine-point block difference scheme (see Figure A.4) uses

$$\nabla^2 u \approx$$
$$\frac{a_1 u_{m,n} + a_2 \left(u_{m+1,n+1} + u_{m-1,n+1} + u_{m-1,n+1} + u_{m-1,n-1} \right) + a_3 \left(u_{m+1,n} + u_{m,n+1} + u_{m-1,n} + u_{m,n-1} \right)}{h^2}$$

(A.10)

where the coefficients equal

$$\begin{bmatrix} a_2 & a_3 & a_2 \\ a_3 & a_1 & a_3 \\ a_2 & a_3 & a_2 \end{bmatrix} = \begin{bmatrix} \dfrac{1-k}{2} & k & \dfrac{1-k}{2} \\ k & -2-2k & k \\ \dfrac{1-k}{2} & k & \dfrac{1-k}{2} \end{bmatrix}$$

k is the weight and takes value in the range 0.6–0.8.
3. Fourth-order difference with respect to space
The nine-point cross-difference scheme (see Figure A.4) uses

$$\nabla^2 u \approx$$
$$\frac{a_1 u_{m,n} + a_3 \left(u_{m+1,n} + u_{m,n+1} + u_{m-1,n} + u_{m,n-1} \right) + a_4 \left(u_{m,n+2} + u_{m,n-2} + u_{m-2,n} + u_{m+2,n} \right)}{h^2}$$

(A.11)

where the coefficients equal

$$
\begin{bmatrix}
 & & a_4 & & \\
 & & a_3 & & \\
a_4 & a_3 & a_1 & a_3 & a_4 \\
 & & a_3 & & \\
 & & a_4 & &
\end{bmatrix}
=
\begin{bmatrix}
 & & \dfrac{-1}{12} & & \\
 & & -\dfrac{4}{3} & & \\
\dfrac{-1}{12} & \dfrac{4}{3} & -5 & -\dfrac{4}{3} & \dfrac{-1}{12} \\
 & & \dfrac{4}{3} & & \\
 & & \dfrac{-1}{12} & &
\end{bmatrix}
$$

The nine-point cross-difference scheme expresses frequency dispersion caused by gridding. However, it is no better than the nine-point block difference scheme when dipping interfaces are involved in the Earth models.

4. First-order difference with respect to density

We use the standard formula for the first-order difference with respect to density as

$$\frac{\partial u}{\partial x} \approx \frac{u_{m+1,n} + u_{m-1,n}}{2h} \tag{A.12}$$

$$\frac{\partial u}{\partial z} \approx \frac{u_{m,n+1} + u_{m,n-1}}{2h} \tag{A.13}$$

$$\frac{\partial \rho}{\partial x} \approx \frac{u_{m+1,n} + u_{m,n}}{h} \tag{A.14}$$

$$\frac{\partial \rho}{\partial z} \approx \frac{u_{m,n+1} + u_{m,n}}{h} \tag{A.15}$$

5. The finite difference equation of wavefield

Substituting Eqns (A8)–(A15) into Eqn (A3) yields

$$u_{m,n,l+1} = -u_{m,n,l-1} + 2u_{m,n} + fa - fb + fx \tag{A.16}$$

where $W(t)$ is the source function,

$$fx = \begin{cases} 0, & s \neq r \\ C^2 \Delta t^2 W(t) \end{cases}$$

For the five-point difference scheme

$$fa = \frac{C^2 \Delta t^2}{h^2} \left[u_{m,n} - 4\left(u_{m+1,n} + u_{m,n+1} + u_{m-1,n} + u_{m,n-1} \right) \right]$$

For the nine-point block difference scheme

$$fa = \frac{C^2 \Delta t^2}{h^2} \left[(-2 - 2k)u_{m,n} + \frac{1-k}{2} \left(u_{m+1,n+1} + u_{m-1,n+1} + u_{m-1,n+1} \right. \right.$$

$$\left. \left. + u_{m-1,n-1} \right) + k \left(u_{m+1,n} + u_{m,n+1} + u_{m-1,n} + u_{m,n-1} \right) \right]$$

For the nine-point cross-difference scheme

$$fa = \frac{C^2 \Delta t^2}{h^2} \left[-5u_{m,n} + \frac{4}{3} \left(u_{m+1,n} + u_{m,n+1} + u_{m-1,n} + u_{m,n-1} \right) - \frac{1}{12} \left(u_{m,n+2} \right. \right.$$

$$\left. \left. + u_{m,n-2} + u_{m-2,n} + u_{m+2,n} \right) \right]$$

For the term of varying density

$$fb = \frac{C^2 \Delta t^2}{2h^2} \left(u_{m+1,n} + u_{m-1,n} \right) \left(\rho_{m+1,n} + \rho_{m,n} \right)$$

A3.3. Conditions of the Stability

In order to keep stable computation, the following conditions must be fitted:

For the block nine – point scheme $\quad \dfrac{C\Delta t}{h} \leq 1 \qquad$ (A.17)

For the cross nine – point scheme $\quad \dfrac{C\Delta t}{h} \leq 0.612 \qquad$ (A.18)

where Δt is the time interval and C is the largest velocity.

A4. PROGRAMMING OF THE FINITE DIFFERENCE METHOD

The program includes three modules that are explained in Table A1 as follows.

A4.1. Finite Difference Modole

It is the main program "fdsm"
Call subroutines: "gridm " and "walet"

TABLE A1   Specification of the Modules

Module	Function	File Name
Finite difference	Wave modeling	fdsm.f
Source wavelet	Generation of wavelet	walet.for
Gridding	Generation of grid mesh	gridm.f

Input files: fdsm.par (contains computation parameters)
            *.mdl (contains the earth model)
            Wavelet data(contains wavelet, optional)
Output files: :*.sei (reflection record)
            *.vsp (VSP record)
            Snapshot files (optional).

A4.2. Gridding

It is the subroutine "gridm"
Format: gridm(nx,ny,nl,dx,fl,c1,xmin,ymin)
Input variables:
            Nx, ny: total horizontal and vertical numbers of nodes
            nl: total number of nodes
            dx = h: size of the grid cell
            fl: file name of the input earth model
Input variables:
            c1: output the earth model for each cell in the grid
            xmin, ymin: start station coordinates

A4.3. Wavelet Options

It is subroutine "walet"
Format: walet(s,nln,dt)
Call for: SUBROUTINE WVLT(N,WA,WB,DF,ISIGN,FPICK,FAZ)
Optional file: "wvlet.par"
Input file: "wvlet.dat" (optional)
Input variables.   nln: length of wavelet (64-128)
            dt: time interval(ms)
            Isign: type of wavelet, has 6 options as shown in Table A2
Input variable.   s(nln): wavelet sequence

TABLE A2   Wavelet Options

Excitation Type	Excitation Parameters		
	isign	**Peak Frequency** *fm*	**Phase** *faz*
Given wavelet	1	Input from "wvlet.dat"	
Minimum phase and exponential decay	5	Specify *fm*	0
Ricker wavelet	4	Specify *fm*	Can be given
Cloutier wavelet	6	Specify *fm*, midband between *fm* ~ 2*fm*	Can be given
Minimum phase and band limit	3	Specify *fm*, midband between *fm* ~ 2*fm*	0

A5.   THE FORTRAN PROGRAM

```
    program fdsm
C===============================
c 2-D Accoutic Equation Modeling
c this program uses finite difference scheme,
c primary version was given by Wencai Yang and Du jianyuan.
c files:
c   fwd.par-------input parameters
c   fdsm.par------input filename and iacc0,ipatt,iden
c   outputfile.par---output file's parameter
c   fout1,fout2------output segy files
c   wvlt.par-----walet paramenters (in fd42.f)
c input data:
c c(i,j) the velocity layer were given by user to use utility file setup
c in this program, it was transform to grid velocity, the densty was
c given in file fdsm.par, the parameters input by paramter file fwd.par,
c the parameter value are given with VB interface.
c Visual Fortran and Visual Basic 6.0 are used.
c test and edited by Wencai Yang 2010/01/17
C================================

   integer iacc0,iden,ipatt
C*  real s(84),c(251001)
   REAL S(84)
   REAL, ALLOCATABLE:: C(:)
C*  dimension pres1(251001),pres2(251001),pres3(251001)
```

```fortran
      REAL, ALLOCATABLE:: PRES1(:),PRES2(:),PRES3(:)
   C* real pp(1000),ck(11)
      REAL CK(11)
      REAL, ALLOCATABLE:: PP(:)
   C* integer pz(150000), px(150000)
      INTEGER, ALLOCATABLE:: PZ(:),PX(:)
      integer dc,nsrty,kc
      character*15 fin,fout1,fout2,parf
      INTEGER(4) NXY,NNXY
      INTEGER ERR_ALLO
   cc    real     xmin,xmax,ymin,ymax,dx,dt,xs,zs,rtl,drtl,xvsp,z0,dz,
gx0,dgx
      data pi/3.1415926/
   C  iter = 0
   C  fmax0 = 0.
   C  fin = 'input.mdl'
   C  fout1 = 'output.sei'
   C  fout2 = 'output.vsp'
   C*********fout1 = fout + .sei, fout2 = fout + .vsp
      open(10,file = 'fdsm.par')
      read(10,*)fin
      read(10,*)fout1,fout2
      read(10,*)parf
      read(10,*)iacc0
      read (10,*)ipatt
      read(10,*)iden
      close(10)
   c open parameter file and load forward parameter
      open(9,file = 'fwd.par')
      read(9,*) xmin,xmax,ymin,ymax
      read(9,*) dx,dt
      read(9,*) nsrty
      read(9,*) xs,zs
      read(9,*) rtl,drtl
      read(9,*) gx0,dgx,ng
      read(9,*) xvsp,z0,dz,nvsp
      read(9,*) istart
      close(9)
      open(10,file = fin)
      read(10,*)xmin,xmax,xscale
      read(10,*)ymin,ymax,yscale
      close(10)
   c caculate forward parameter
```

```fortran
      nx = ifix((xmax − xmin)/dx) + 1
      ny = ifix((ymax − ymin)/dx) + 1
      n1 = ifix(zs/dx) + 1
      nss = ifix(xs/dx) + 1
      if(nss.le.2) nss = 3
      if(n1.eq.ny) n1 = ny − 1
      nr = ifix(xvsp/dx) + 1
      nz0 = ifix(z0/dx) + 1
      istop = ifix(rtl*1000./dt)
      hh = dx
      idt = ifix(drtl/dt)
      if(idt.eq.0) then
      idt = 1
      drtl = dt
      endif
      msam = ifix(rtl*1000./drtl)
      kz = 0
      kx = 0
      open(10,file = parf)
      write(10,*) gx0,dgx,ng
      write(10,*) msam,drtl
      write(10,*) xvsp,z0,dz,nvsp
      close(10)
C open files
      if(iacc0.eq.0) then
        open(19,file = fout1,form = 'binary',access = 'direct',recl = 4)
        open(20,file = fout2,form = 'binary',access = 'direct',recl = 4)
      ELSE
        OPEN(19,FILE = FOUT1)
        OPEN(20,FILE = FOUT2)
      ENDIF

      call walet(s,84,dt)
c griden a Earth model
      nxy = nx*ny
      If(.NOT.ALLOCATED(C)) ALLOCATE(C(NXY),STAT = ERR_ALLO)
      IF(ERR_ALLO.NE.0) PRINT *, "C Allocation error"
      PRINT *, "GRIDMING......"
      call gridm(nx,ny,nxy,dx,fin,c,xmin,ymin)
      PRINT *,"GRIDMING BE FINISH."
      cmax = 0.
      cmin = 9999.
      do 19 i = 1,nxy
C  if(c(i).eq.0.) write(*,*) i
```

```
      cmax = max(cmax,c(i))
      cmin = min(cmin,c(i))
 19   continue
      dltc = (cmax - cmin + 1)/10
      do 21 i = 1,11
      ck(i) = cmin + (i - 1)*dltc
 21   continue
c forward finite difference scheme
      jcb = 2
      jce = nx - 1
C*  NNXY = (NX + 1)*(NY + 1)
      If(.NOT.ALLOCATED(PRES1)) ALLOCATE (PRES1(NXY),STAT = ERR_ALLO)
      IF(ERR_ALLO.NE.0) PRINT *, "PRES1 Allocation error"
     If(.NOT.ALLOCATED(PRES2)) ALLOCATE (PRES2(NXY),STAT = ERR_ALLO)
      IF(ERR_ALLO.NE.0) PRINT *,"PRES2 Allocation error"
      IF(.NOT.ALLOCATED(PRES3)) ALLOCATE (PRES3(NXY),STAT = ERR_ALLO)
      IF(ERR_ALLO.NE.0) PRINT *,"PRES3 Allocation error"
        do 190 i = 1,nxy
      pres1(i) = 0.
      pres2(i) = 0.
      pres3(i) = 0.
 190  continue
      If(.NOT.ALLOCATED(PZ)) ALLOCATE(PZ(MSAM*NG),STAT = ERR_ALLO)
      IF(ERR_ALLO.NE.0) PRINT *,"PZ Allocation error"
      do 191 i = 1,msam*ng
      pz(i) = 0
 191  continue
      If(.NOT.ALLOCATED(PX)) ALLOCATE(PX(MSAM*NVSP),STAT = ERR_ALLO)
      IF(ERR_ALLO.NE.0) PRINT *,"PX Allocation error"
      do 192 i = 1,msam*nvsp
      px(i) = 0
 192  continue
      kkkk = (nr - 1)*ny
      if(ipatt.eq.0) irecv = 2
      if(ipatt.eq.1) irecv = ny - 2
      If(.NOT.ALLOCATED(PP)) ALLOCATE(PP(NX),STAT = ERR_ALLO)
      IF(ERR_ALLO.NE.0) PRINT *,"PP Allocation error"
      PRINT *,"CALCUTATION IS BEGINNIG......"
      do 200 inct = istart,istop
      dc = int(cmax*dt*inct/(dx*1000.))
      irb = n1 - dc - 2
      ire = n1 + dc + 4
      if(nsrty.eq.1) then
      jcb = nss - dc - 2
```

```
    jce = nss + dc + 2
    endif
    if(nsrty.eq.0) then
    jcb = 2
    jce = nx - 1
    irb = 2
    endif
    if(irb.lt.2) irb = 2
    if(ire.gt.ny - 1) ire = ny - 1
    if(jcb.lt.2) jcb = 2
    if(jce.gt.nx - 1) jce = nx - 1
      do 201 i = 1,nx
201   pp(i) = 0.
    if(inct.gt.79) goto 204
    if(nsrty.eq.0) goto 202
    ll = nss
    pp(ll) = s(inct + 4)
    pp(ll - 1) = 0.7*s(inct + 2)
    pp(ll + 1) = pp(ll - 1)
    pp(ll - 2) = 0.3*s(inct)
    pp(ll + 2) = pp(ll - 2)
    goto 204
202   do 203 k = 1,nx
    kc = inct + 5
    if(kc.lt.1) goto 203
    if(kc.gt.79) goto 203
    pp(k) = s(kc)
203   continue
204   do 210 j = 2,nx - 1
    kk = (j - 1)*ny + 1
    if(n1.eq.1) then
    pres3(kk) = pp(j)*c(kk)*dt*dt*c(kk)/1000000.
    else
    pres3(kk) = 0.
    endif
210   continue
c set left and right sides boundary condation
    do 255 i = 1,ny
    cnt1 = c(i)*dt/(hh*1000.)
    knx = (nx - 1)*ny + i
    cnt2 = c(knx)*dt/(hh*1000.)
    k2 = ny + i
    k3 = 2*ny + i
    knx1 = (nx - 2)*ny + i
```

```
      knx2 = (nx − 3)*ny + i
      pres3(i) = pres2(i) + pres2(k2) − pres1(k2)+
     1 cnt1*(pres2(k2) − pres2(i) − pres1(k3) + pres1(k2))
      pres3(knx) = pres2(knx) + pres2(knx1) − pres1(knx1)
     1 −cnt2*(pres2(knx) − pres2(knx1) − pres1(knx1) + pres1(knx2))
 255  continue
c set bottom side boundary condation
      do 261 j = 2,nx − 1
      kj = j*ny
      cnt1 = c(kj)*dt/(hh*1000.)
      pres3(kj) = pres2(kj) + pres2(kj − 1) − pres1(kj − 1)−
     1 cnt1*(pres2(kj) − pres2(kj − 1) − pres1(kj − 1) + pres1(kj − 2))
 261  continue
c main scheme
      do 250 ir = irb,ire
      do 250 jc = jcb,jce
      l1 = (jc − 1)*ny + ir
      l2 = jc*ny + ir
      l3 = (jc − 2)*ny + ir
      l4 = l1 + 1
      l5 = l1 − 1
      cnt = c(l1)*c(l1)*dt*dt/(hh*hh*1000000.)
      if(ir.eq.n1) then
      l6 = (jc − 1)*ny + n1
      fx = c(l6)*c(l6)*dt*dt*pp(jc)/1000000.
      else
      fx = 0.
      endif
      if(iden.eq.1) then
      sqc = sqrt(sqrt(c(l1)))
      rho1 = 0.5*sqrt(sqrt(c(l4)))/sqc
      rho2 = 0.5*sqrt(sqrt(c(l2)))/sqc
      cxzt = 2. − 4.*cnt
      cxz0t = cnt*(0.5 + rho1)
      cx0zt = cnt*(0.5 + rho2)
      cxz1t = cnt*(1.5 − rho1)
      cx1zt = cnt*(1.5 − rho2)
      pres3(l1) = −pres1(l1) + cxzt*pres2(l1) + cx1zt*pres2(l2)
     1 +cx0zt*pres2(l3) + cxz1t*pres2(l4) + cxz0t*pres2(l5) + fx
      else
      pres3(l1) = −pres1(l1) + 2*pres2(l1) + cnt*(pres2(l2)+
     1 pres2(l3) − 4.*pres2(l1) + pres2(l4) + pres2(l5)) + fx
      endif
 250  continue
```

```
c check overflow and recived reflection wave
  sum = 0.
  if(mod(inct,idt).eq.0.and.inct.ge.40) then
  kx = kx + 1
  do 310 k = 1,ng
C*  modify below 3 columns-----20080901
C   kl = int((gx0 + (k − 1)*dgx)/dx)
C   if(kl.gt.nx − 1) goto 310
C   km = (kl − 1)*ny + irecv
C*  new columns-----20080901
  KL = INT((GX0 + (K − 1)*DGX)/DX) + 1
  IF(KL.GT.NX)GOTO 310
  KM = KL*NY + IRECV
  xx = pres3(km)
  if(abs(xx).lt.10e − 18) xx = 0.
  kmm = (k − 1)*msam + kx
  pz(kmm) = int(xx*10000.)
  sum = sum + xx
310  continue
c received VSP signal
  kz = kz + 1
  do 320 k = 1,nvsp
  kn = ifix((z0 + (k − 1)*dz)/dx)
  kzz = (k − 1)*msam + kz
  krr = kkkk + kn
  xx = pres3(krr)
  px(kzz) = ifix(xx*1000.)
320  continue
  endif
  if(sum.gt.1.e9) goto 808
c display snapshot in some time length(Deleted)

c next iteration
208  do 290 i = 1,nxy
  pres1(i) = pres2(i)
  pres2(i) = pres3(i)
  pres3(i) = 0.
290  continue
200  continue
c save result file
808  continue

C********* parf = fout1 + .par
C   open(10,file = parf)
```

```
C  write(10,*) gx0,dgx,ng
C  write(10,*) msam,drt1
C  write(10,*) xvsp,z0,dz,nvsp
C  close(10)
C write files
   nn = msam*ng
   mm = msam*nvsp
   kmax = 0
   do 302 K = 1,NG
   DO 302 J = 1,MSAM
   I = (K − 1)*MSAM + J
   IF(NSRTY.EQ.0) THEN
   IF(J.LE.50) PZ(I) = 0
   ENDIF
   kmax = max0(kmax,abs(pz(i)))
302  continue
   kmax = max0(kmax,1)
   lmax = 0
   do 303 i = 1,mm
   lmax = max0(lmax,abs(px(i)))
303  continue
   lmax = max0(lmax,1)
   do 304 i = 1,nn
   pz(i) = int(pz(i)*7000/kmax)
304  continue
   do 305 i = 1,mm
   px(i) = int(px(i)*3000/lmax)
305  continue
   if(iacc0.eq.0) then
C  open(9,file = fout1,form = 'binary',access = 'direct',recl = 4)
   do 306 i = 1,nn
   write(19,rec = i) pz(i)
306  continue
   close(19)
C  open(20,file = fout2,form = 'binary',access = 'direct',recl = 4)
   do 307 i = 1,mm
   write(20,rec = i) px(i)
307  continue
   close(20)
   else
C  open(9,file = fout1)
   do 616 i = 1,nn
   write(19,*) pz(i)
616  continue
```

```
  close(19)
C  open(10,file = fout2)
  do 617 i = 1,mm
  write(20,*) px(i)
617  continue
  close(20)
  DEALLOCATE(PX,PZ)
  DEALLOCATE(PP)
  DEALLOCATE(C,PRES1,PRES2,PRES3)
  endif
  PRINT *,"CALCUTATION FINISH"
C================================================
c format gramma
104  format(i1)
101  format(i4)
102  format(f6.3)
103  format(f6.0)
606  format(12i6)
777  format(f9.3)
C999  call text_mode()
  stop
  end
C--------------------------------------------------------
C subroutine follows
  subroutine gridm(nx,ny,nl,dx,fl,cl,xmin,ymin)
C=======================================
c this subprogram converts primary Earth model to Grid
c Earth model, in this program.
c==================By Du Jianyuan======Feb. 1993=========
  integer nk,m1,ic0
  real vel
C  integer mr(256)
CC   real c2(400,64),h(400,64)
  REAL C2(NX,64),H(NX,64)
CC  REAL, ALLOCATABLE:: C1(:)
  INTEGER(4) JI
  real c1(nl)
  real x(64,64),y(64,64)
  character fl*15
  nh = 1
  do 10 i = 1,nx
  do 10 j = 0,64
  c2(i,j) = 0.
10  h(i,j) = 0.
```

```
c read data
  open(11,file = fl)
  read(11,*) x0,x1,xscale
  read(11,*) y0,y1,yscale
  read(11,*) nf
  do 20 i = 1,nf
  read(11,*) ic0,nk,vel
  read(11,111) x(i,1),y(i,1)
  if(i.ne.1) then
  if(x(i,1).lt.x(i − 1,kn1)) then
  nh = nh + 1
  endif
  endif
  do 30 j = 2,nk
  read(11,111) x(i,j),y(i,j)
  i0 = int((x(i,j − 1) − x0)/dx) + 1
  ie = int((x(i,j) − x0)/dx) + 1
  x2 = x(i,j) − x(i,j − 1)
  y2 = y(i,j) − y(i,j − 1)
  if(x2.eq.0.) goto 30
  al = y2/x2
  do 40 k = i0,ie
  xk = x0 + (k − 1)*dx
  yk = al*(xk − x(i,j − 1)) + y(i,j − 1)
  kk = int((xk − xmin)/dx) + 1
  if(kk.ge.1) then
  c2(kk,nh) = vel
  h(kk,nh) = yk
  endif
40  continue
30  continue
  read(11,*) iii,jjj
  kn1 = nk
20  continue
111  format(2f8.0)
  close(11)
c according to layer to setup data
  do 50 i = 1,nx
  nn = 1
  do 60 j = 1,nh
  if(h(i,j).gt.0.) then
  h(i,nn) = h(i,j)
  c2(i,nn) = c2(i,j)
  nn = nn + 1
```

```
   endif
60  continue
50  continue
c griden depth
CC  ALLOCATE (C1(NL),STAT = ERR_ALLOC)
CC  IF(ERR_ALLOC.NE.0) PRINT *,"ALLOCATION ERROR"
   do 70 j = 2,ny
   yc = ymin + dx*(j − 1)
   do 80 i = 1,nx
   ji = (i − 1)*ny + j
   do 90 k = 1,nh
   if(yc.gt.h(i,k − 1).and.yc.le.h(i,k)) then
   c1(ji) = c2(i,k)
   goto 80
   endif
90  continue
80  continue
70  continue
   do 100 i = 1,nx
   ji = (i − 1)*ny + 1
   c1(ji) = c1(ji + 1)
100  continue
   do 110 i = 1,nx
   ji = i*ny
   c1(ji) = c1(ji − 1)
110  continue
   return
   stop
   end
C-----------------------------------------------------------------
subroutine walet(s,nln,dt)
C PRODUCE A WAVELET FOR SYNTHETIC SEISMOGRAPHY
   real wb(1024),wa(1024),s(nln)
   INTEGER len,isign
   REAL fpick,phz,df
cc  character fl*15,cv*15,string*60,nul*1,cv1*8
cc  nul = char(0)
   phz = 0.
   do 111 i = 1,nln
   s(i) = 0.
111  continue
C FORM WAVELET IN THE FREQUENCY DOMAIN
   open(10,file = 'WVLT.PAR')
c  read(10,*)dt
```

```
  read(10,*)len
  read(10,*)isign
  read(10,*)fpick
  read(10,*)phz
  close(10)
c*******usually pick len = 1024
  df = 1000./dt/1024.
  call wvlt(1024,wa,wb,df,isign,fpick,phz)
C MAKE A WAVELET WITH 80 SAMPLES
  call fastf(wa,wb,1024,1)
  do 500 k = 1,39
  s(k) = wa(40 - k)
  IF (S(K).LT.0.) S(K) = 0.618*S(K)
500  continue
  do 511 k = 40,79
  s(k) = wa(1024 - k + 40)
  IF (S(K).LT.0.) S(K) = 0.618*S(K)
511  continue
  s(1) = 0.5*s(2)
  s(80) = 0.5*s(79)
  wm = 0.
  do 510 k = 1,80
  awb = abs(s(k))
  if(awb.gt.wm) wm = awb
510  continue
  do 520 k = 1,80
  s(k) = s(k)/wm
520  continue
C SAVE WALET DATA IN S(K) TO FILE 'WALET.DAT' FOR DISPLAY AFTERWORD
  OPEN(10,FILE = 'WALET.DAT')
  DO 521 K = 1,80
521 WRITE(10,555) S(K)
  CLOSE(10)
555 FORMAT(10F8.5)
  return
  stop
  end
C-----------------------------------------------------------------
  SUBROUTINE WVLT(N,WA,WB,DF,ISIGN,FPICK,FAZ)
C PROVIDE 6 KINDS OF SIMPLE WAVELET FOR SYNTHETIC TRACES IN FREQENCY
C DOMAIN WITH REAL PART WA(K) AND IMAGINARY PART WB(K) OF LENGTH N.
C  ISIGN = 1 : FOR GIVEN WAVELET OF LENGTH EQUAL TO N/8
C   =2 : FOR ZERO PHASE PULSE WAVELET
C   =3 : MINIMUN PHASE WAVELET OF BAND-LIMIT AS ISIGN = 6
```

```
C    PARAMETER FAZ = PHAZE(I)
C    =4 : PHASE = FAZ RICKER WAVELET WITH MAXIMUN FPICK
C    =5 : MINIMUM PHASE WAVELET = EXP(−.25*K)*COS(2*PI*K/4)
C    =6 : Cloutier WAVELET WITH MAXIMUN FPICK USED FOR A SIGNITURE OF
VIBRATOR
C   FPICK : PICK FREQUENCY OF WAVELET
C   FAZ  : CONSTANT PHASE OF THE WAVELET
C   OUTPUT IS THE SPECTRUM OF WAVELET IN WA AND WB
C================ YANG, 1985,11, =====================
    REAL WA(N),WB(N),WORK(6),PHAZE(2048)
    DATA WORK/8.,25.,60.,80.,0.9,0./,FMAX/25./
    IF(FPICK.GE.1.) FMAX = FPICK
    F2 = 389.8484*FLOAT(IP)/16.
    N2 = N/2
    N1 = N2 + 1
    DO 5 I = 1,N
    WA(I) = 0.
 5  WB(I) = 0.
    M2 = N − 2
    N8 = N/8
    GOTO(110,120,130,140,150,160), ISIGN
C MINIMUM-PHASE WAVELET
150 NST = 50
    DO 15 I = NST,N2
    T = FLOAT(I − NST)
    WA(I) = EXP(−.25*T)*COS(3.14159265*T/5.)
 15 CONTINUE
    WA(NST − 1) = 0.5*WA(NST)
    DO 17 I = N2,N − 16
    T = FLOAT(I − N2)
    WA(I) = EXP(−.25*T)*COS(3.14159265*T/8.)
 17 CONTINUE
    WA(N2) = 0.5*(WA(N2 + 1) + WA(N2 − 1))
    GOTO 20
110  OPEN (UNIT = 10,FILE = 'WVLT.DAT',STATUS = 'OLD')
    READ(10,1)(WA(I),I = 1,N8)
 1  FORMAT(10F6.4)
    CLOSE(UNIT = 10)
 20 WA(N) = 0.5*(WA(1) + WA(N − 1))
    CALL FASTF(WA,WB,N,−1)
170 DO 30 K = 1,M2,1
    A1 = WA(K + 1) − WA(K)
    WA(K) = WA(K) + 0.5*A1 − 0.125*(WA(K + 2) − WA(K + 1) − A1)
    B1 = WB(K + 1) − WB(K)
```

```
 30 WB(K) = WB(K) + 0.5*B1 − .125*(WB(K + 2) − WB(K + 1) − B1)
    WA(N − 1) = WA(N − 1) + .5*(WA(N) − WA(N − 1)) − .125*(WA(1) + WA(N − 1) −
2.*WA(N))
    WB(N − 1) = WB(N − 1) + .5*(WB(N) − WB(N − 1)) − .125*(WB(1) + WB(N − 1) −
2.*WB(N))
    WA(N) = WA(N) + .5*(WA(1) − WA(N)) − .125*(WA(2) + WA(N) − 2.*WA(1))
    WB(N) = WB(N) + .5*(WB(1) − WB(N)) − .125*(WB(2) + WB(N) − 2.*WB(1))
    RETURN
C FOR SMOOTH INPULS
120 DO 50 K = 1,N1
    WA(K) = 0.5*(1. − COS(3.14159265*(N1 − K + 1)/N1))
    WB(K) = 0.
 50 CONTINUE
    DO 51 K = N1 + 1,N
    WA(K) = WA(N − K + 2)
 51 WB(K) = WB(N − K + 2)
999 RETURN
140 CONTINUE
C OPTION 4: THE LONG-RANGE RICKER WAVELET APPROXIMATION
    PHZ = FAZ/57.29578
    DO 149 K = 1,N1
    F2 = ((K − 1)*DF/FMAX)**2
    WA(K) = 1.128379*(K − 1)*F2*EXP(−F2)
    WB(K) = −WA(K)*SIN(PHZ)
149 WA(K) = WA(K)*COS(PHZ)
    DO 147 K = N1 + 1,N
    WA(K) = WA(N − K + 2)
147 WB(K) = WB(N − K + 2)
    RETURN
C MINIMUM PHASE BAND-LIMIT WAVELET, PARAMETER AS ISIGN = 6
130 N4 = INT(FPICK/DF)
C
    A1 = −42./57.29578
    CALL LINTP(0.,A1,WB,N4)
    DO 300 J = 1,N4
300 PHAZE(J) = WB(J)
    N3 = N2 − N4
    CALL LINTP(A1,0.,WB,N3)
    DO 310 J = N4 + 1,N4 + N3
310 PHAZE(J) = WB(J)
    PHAZE(N1) = 0.
    DO 315 J = 2,N2
```

```
315 PHAZE(J) = 0.5*PHAZE(J) + 0.25*(PHAZE(J − 1) + PHAZE(J + 1))
    DO 320 J = N1 + 1,N
 320 PHAZE(J) = −PHAZE(N − J + 2)
    GOTO 160
 C
 C FOR CONSTANT FHASE WAVELET PHZ = CONSTANT, PARAMETERS ARE
 C  F1, FPICK, F3 AT 0.8PICK, F4, AF3 = 0.8, PHZ, DF
 C160 OPEN (UNIT = 10,FILE = 'WVLT.DAT',STATUS = 'OLD')
 C  READ(10,*)(WORK(I),I = 1,6)
 C  CLOSE (UNIT = 10)
 160 IF(ISIGN.EQ.6) PHZ = FAZ
 C  WRITE(*,*) PHZ
    WORK(2) = FMAX
    WORK(3) = 2.*FMAX
    WORK(4) = WORK(2)*3.
    PHZ = PHZ/57.29578
 330 LF1 = INT(WORK(1)/DF + .5)
    LF2 = INT(WORK(2)/DF + .5)
    LF3 = INT(WORK(3)/DF + .5)
    LF4 = INT(WORK(4)/DF + 0.5)
 C
    L1 = LF2 − LF1
    CALL LINTP(0.,1.,WB,L1)
    DO 910 L = LF1,LF2 − 1
    WA(L) = WB(L − LF1 + 1)
 910 WA(N − L + 2) = WA(L)
    L2 = LF3 − LF2
    CALL LINTP(1.,WORK(5),WB,L2)
    DO 920 L = LF2,LF3 − 1
    WA(L) = WB(L − LF2 + 1)
 920 WA(N − L + 2) = WA(L)
    L3 = LF4 − LF3
    CALL LINTP(WORK(5),0.,WB,L3)
    DO 930 L = LF3,LF4 − 1
    WA(L) = WB(L − LF3 + 1)
 930 WA(N − L + 2) = WA(L)
    DO 933 I = 2,N − 1
 933 WA(I) = .25*(WA(I + 1) + WA(I − 1)) + WA(I)*.5
    DO 940 I = 1,N
    IF (ISIGN.EQ.6) THEN
    WB(I) = −WA(I)*SIN(PHZ)
    IF(I.GT.N/2 + 1)WB(I) = −WB(I)
    WA(I) = WA(I)*COS(PHZ)
```

```
    ELSE
      WB(I) = -WA(I)*SIN(PHAZE(I))
      IF(I.GT.N/2 + 1)WB(I) = -WB(I)
      WA(I) = WA(I)*COS(PHAZE(I))
    END IF
940 CONTINUE
    WB(N/2 + 1) = 0.
    END
C------------------------------------------------------------
    SUBROUTINE LINTP(Y1,Y2,Y,K)
    REAL Y(K)
    IF (K.LE.1) GOTO 20
    Y(1) = Y1
    KB = 1
    DO 10 J = 2,K
    Y(J) = (Y1*(K - KB) + Y2*KB)/FLOAT(K)
10  KB = KB + 1
20  RETURN
    END
C------------------------------------------------------------
    SUBROUTINE FASTF (FR,FI,N,ISIGH)
C   FAST FT WITHOUT COMPLEX NUMBER CALCULATION
C   FR & FI ARE THE REAL & IMAGINARY PARTS RESPECTIVELY
C   N IS THE LENGTH OF VECTORS FR OR FI, POWER OF 2
C   ISIGH = 1 FOR INVERSE FT
C    =-1 FOR FORWARD FT
    REAL FR(N),FI(N),GR,GI,ER,RI,EU,EZ
C TEST IF N = POWER OF 2
    MQ = 4
    DO 50 K = 1,12
    IF(N.EQ.MQ) GOTO 59
    MQ = MQ*2
50  CONTINUE
    WRITE(*,54)
54  FORMAT(' WRONG DATA: N IS NOT POWER OF 2 !!!')
    STOP
C
59  M = 0
    KD = N
1   KD = KD/2
    M = M + 1
    IF (KD.GE.2) GOTO 1
    ND2 = N/2
    NM1 = N - 1
```

```
      L = 1
      DO 4 K = 1,NM1
      IF(K.GE.L) GOTO 2
      GR = FR(L)
      GI = FI(L)
      FR(L) = FR(K)
      FI(L) = FI(K)
      FR(K) = GR
      FI(K) = GI
2     NND2 = ND2
3     IF(NND2.GE.L) GOTO 4
      L = L - NND2
      NND2 = NND2/2
      GOTO 3
4     L = L + NND2
      PI = FLOAT(ISIGH)*3.14159265
      DO 6 J = 1,M
      NJ = 2**J
      NJD2 = NJ/2
      EU = 1.
      EZ = 0.
      ER = COS(PI/NJD2)
      EI = SIN(PI/NJD2)
      DO 6 IT = 1,NJD2
      DO 5 IW = IT,N,NJ
      IWJ = IW + NJD2
      GR = FR(IWJ)*EU - FI(IWJ)*EZ
      GI = FI(IWJ)*EU + FR(IWJ)*EZ
      FR(IWJ) = FR(IW) - GR
      FI(IWJ) = FI(IW) - GI
      FR(IW) = FR(IW) + GR
5     FI(IW) = FI(IW) + GI
      SEU = EU
      EU = SEU*ER - EZ*EI
6     EZ = EZ*ER + SEU*EI
      IF(ISIGH.EQ.-1) GOTO 70
      FF = FLOAT(N)
      DO 7 KK = 1,N,1
      FR(KK) = FR(KK)/FF
7     FI(KK) = FI(KK)/FF
70    RETURN
      END
```

A6. INSTRUCTION MANUAL FOR THE PROGRAM

Working on a microcomputer with the program, one may follow the instructive steps as follows.

1. Copy programs in to "/yfdsm" and active the programs.
2. Input data, set the parameter file "FDSM.par". One can uses a shot program as

```
open(10,file = 'fdsm.par')
read(10,*)fin: give filename of the earth model(input.mdl)
read(10,*)fout1,fout2: define output filenames for both reflection
and VSP(*.sei)and (*.vsp) respectively
read(10,*)parf: define filename of finite difference (fwd.par)
read(10,*)iacc0 = 1: define output data format, sequential = 1,
direct = 0 (no good for converse to SGY format)
read (10,*)ipatt: define observation mode, reflection and VSP = 0,
cross-hole = 1
read(10,*)iden: specify whether density is variable or not, no = 0,
yes = 1
close(10)
```

3. Create the earth model

To create a earth model into a file called "input.mdl", one must follow the fixed format as shown as follows. For example, a two-layer model is expressed as

```
0.0000000E+00 1000.000 100.0000 (1 km width, 10 marks for display),
0.0000000E+00 1000.000 100.0000 (1 km depth, 10 marks for display),
        2       (total layer number)
    1   2 2000.000 layer 1, defining points, velocity
     0. 200.   (x,y) of the first defining point
  1000. 200.   (x,y) of the second defining point
     0   0 end of defining the first layer
     2   2 2100.000 layer 2, defining points, velocity
     0. 1000.  (x,y) of the first defining point
  1000. 1000.  (x,y) of the second defining point
     0   0 end of defining the second layer
```

4. Set finite difference parameters file "FWD.par"

Because this file is erased after computation, it must be saved in another file block as \save, and copied to\YFDSM. The format of the file is

```
0  1000  0  1000   section size (m)
2  .1   cell size (m) and time difference dt(ms)
0  Mode, stacked modeling = 0, prestack common shot = 1
1  200  0    prestack shot location(xs,xz)
1  1    record length(s) and sampling rate (ms)
5  10   100  start point x, channel dx & number ns
800  100   10   80 VSP hole x,z1,dz,ns
```

5. Set wave parameters in file "WVLT.par"

1024	Length must be powers of 2
6	Wavelet type, as Isign = (1—6)
30	Peak frequency fm
1024	Phase shift

6. Execution of the program

Execution of FDSM.exe produces output files.

7. Converting the output data files to SEGY files and display.

8. Modifying the Earth model for another job.

Be careful to copy the file "input.mdl" to working block from/save/.

Figures A5—A8 show the output from a five-layer model generated by the program. The third layer is a thin layer and assumed to be the exploration target. The Earth model contains 500×500 cells with its size of 2 m. The peak frequency is 80 Hz; Ricker wavelet (isign = 4) is used. The output consists of stacked reflection profiles with four different schemes or parameters. A graduate student, Mr Zeng Xian-Zi, presented these results in 2008. Comparison of the results yields some understanding about the finite difference procedure. The nine-point cross difference scheme shows the best image in this reflection seismic modeling job.

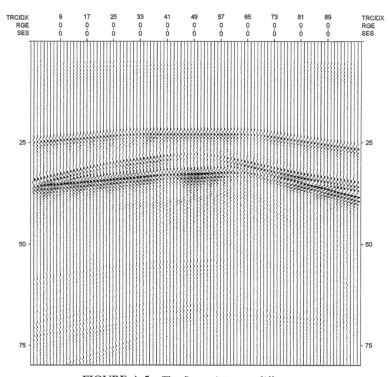

FIGURE A.5   The five-point cross-difference.

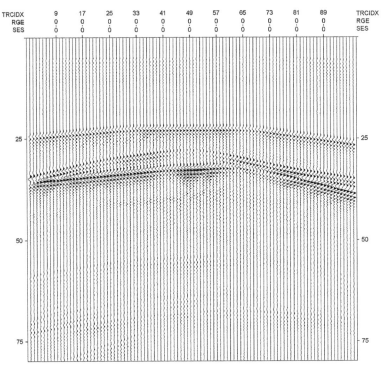

FIGURE A.6   Nine-point block, $k = 0.6$.

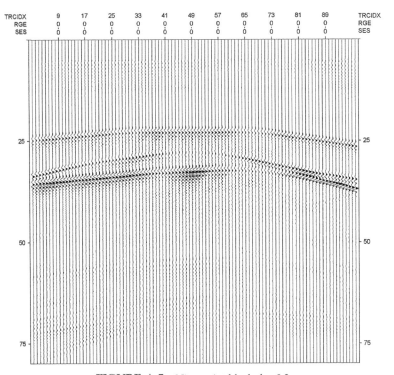

FIGURE A.7   Nine-point block, $k = 0.3$.

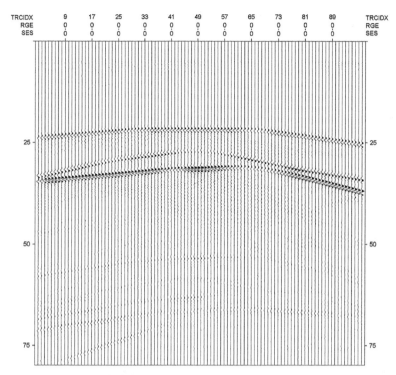

FIGURE A.8   Nine-point cross-difference.

References

Aarts, E., Korst, J., 1989. Simulated Annealing and Boltzmann Machines. John Wiley & Sons, New York.

Aki, K., Richards, P.G., 1980. Quantitative Seismology: Theory and Methods, vol. 1. W. H. Freeman Co, San Francisco.

Aubin, J.P., 1979. Applied Functional Analysis. John Wiley & Sons Inc, New York.

Backus, G.E., Gilbert, J.F., 1967. Numerical application of a formulism for geophysical inverse problems. Geophys. J. R. Astr. Soc. 13, 247–276.

Backus, G.E., Gilbert, J.F., 1970. Uniqueness in inversion of inaccurate gross earth data. Phil. Trans. R. Soc. Lond. A 266, 123–192.

Bender, C.M., Orszag, S.A., 1988. Advanced Mathematical Methods for Scientists and Engineers. McGraw-Will, New York.

Berkout, A.J., 1985. Seismic Migration, Imaging of Acoustic Energy by Wavefield Extrapolation. Elsevier.

Berryman, J.G., 1995. Mixture Theories for Rock Properties. In: Ahrens, T.J. (Ed.), Rock Physics and Phase Relations. American Geophysical Union, pp. 205–208.

Berryman, J.G., 2007. Seismic waves in rocks with fluids and fractures. Geophys. J. Int. 171, 954–972.

Beyer, W.H., 1981. CRC Standard Mathematical Tables. CRC Press Inc, Florida.

Beylkin, G., Rurridge, R., 1990. Linearlized inverse problems in acoustic and elasticity. Wave Motion 12, 15–52.

Biot, M.A., 1956. Theory of propagation of elastic waves in a fluid-saturated solid. J. Acoust. Soc. Am. 28, 168–191.

Biot, M.A., 1962. Mechanics of propagation of elastic waves in a fluid-saturated solid. J. Appl. Phys. 33, 1482–1498.

Boore, D.M., 1972. Finite element methods for seismic wave propagation in heterogeneous material. Methods Comput. Phys. 11.

Bosch, M., Mukerji, T., Gonzalez, E.F., 2010. Seismic inversion for reservoir properties combining statistical rock physics and geostatistics: a review. Geophysics 75 (5), 165–166.

Budak, B.M., Samarskii, A.A.S., Tikhonov, A.N., 1964. A Collection of Problems on Mathematical Physics. Pergamon Press, Oxford.

Cerveny, V., 2001. Seismic Ray Theory. Cambridge University Press, Cambridge.

Christenson, R.M., 1982. Theory of Viscoelasticity. Academic Press, New York.

Claerbout, J.F., 1985. Imaging the Earth's Interior. Blackwell Scientific Pub.

Cohen, J.K., Bleistein, N., 1979. Velocity inversion procedure for acoustic waves. Geophysics 44, 1497–1511.

Cohen, J.K., Bleistein, N., 1986. Three-dimensional born inversion with an arbitrary reference. Geophysics 51, 1552–1558.

Davis, L.D., 1991. Handbook of Genetic Algorithms. Van Nostrand Reinhold.

Eringen, A.C., Suhubi, E.S., 1975. Elastodynamics. Academic Press, New York.

Ewing, M., Jardetzky, W.S., Press, F., 1957. Elastic Waves in Layered Media. McGraw Hill, New York.

Freidman, B., 1965. Principles and Techniques of Applied Mathematics. McGraw Hill, New York.

Fung, Y.C., 1977. A First Course in Continuum Mechanics. John Wiley & Sons.

Futterman, W.I., 1962. Dispersive body waves. J. Geophys. Res. 67, 5279−5291.

Garnir, H.G. (Ed.), 1980. Singularities in Boundary Value Problems. D. Reidel Pub. Co, Amsterdam.

Gassmann, F., 1951. Elastic waves through a packing of spheres. Geophysics 16, 673−685, 18, 269.

Geldart, L.P., Sheriff, R.E., 2004. Problems in Exploration Seismology and Their Solutions. SEG Pub. Ser. 14.

Holland, J.H., 1975. Computation in Natural and Artificial Systems. University of Michigan Press, Ann Arbor.

Hubral, P., Schleicher, J., Tyge, L.M., 1992. Three-dimensional paraxial ray properties: part 1 basic relations. J. Seism. Explor. 1, 265−279.

Hutson, V., Pym, J.S., 1980. Applications of Functional Analysis and Operator Theory. Academic Press, London.

James, D.E. (Ed.), 1989. Encyclopedia of Solid Earth Geophysics. Van Nostrand Reinhold Com, New York.

Ji, S.C., Wang, Q., Xia, B., 2002. Handbook of Seismic Properties of Minerals, Rocks and Ores. Polytechnic International Press, Montreal.

Ji, S.C., Wang, Q., Marcotte, D., et al., 2007. P wave velocities, anisotropy and hysteresis in ultrahigh-pressure metamorphic rocks as a function of confining pressure. J. Geophys. Res. 112, B09204. http://dx.doi.org/10.1029/2006JB004867.

Kaplan, D., Glass, L., 1995. Understanding Nonlinear Dynamics. Springer-Verlag, New York.

Kennett, B.L.N., 1983. Seismic Wave Propagation in Stratified Media. Cambridge University Press, London.

Kuster, G., Toksöz, M.N., 1974. Velocity and attenuation of seismic waves in low-phase media. Part I: Theoretical formulations. Geophysics 39, 587−606.

Liang-Jie, H., Wen-Cai, Y., 1991a. Inversion of the acoustic wave equation by inverse scattering. Chin. J. Geophys. 34 (2), 331−341. Allerton Press Inc., New York.

Liang-Jie, H., Wen-Cai, Y., 1991b. Approximation method of the inversion of the acoustic wave equation by inverse scattering. Chin. J. Geophys. 34 (3), 489−498. Allerton Press Inc., New York.

Lines, L.R., Treitel, S., 1983. Digital filtering with the second moment norm. Geophysics 48, 505−514.

Lliboutry, L., 1999. Quantitative Geophysics and Geology. Springer.

Marozov, V.A., 1984. Methods for Solving Incorrectly Posed Problems. Springer-Verlag, New York.

Maslov, V.P., 1965. Theory of Perturbation and Asymptotic Methods (in Russian). Izd MGU, Moscow.

Mavko, G., Tapan, B., Muker, J., Dvorkin, J., 2009. Rock Physics Handbook. Cambridge University Press, Cambridge.

Menke, W., 1984. Geophysical Data Analysis: Discrete Inverse Theory. Academic Press, London.

Metropolis, M., Rosenbluth, A., 1953. Equation of state calculations by fast computing machines. J. Chem. Phys. 21, 1087−1092.

Mrinal, S., Stoffa, P.L., 1995. Global Optimization Methods in Geophysical Inversion. Elsevier.

Murphy, W.F., Winkler, K.W., Kleinberg, R.L., 1986. Acoustic relaxation in sedimentary rocks: dependence on grain contacts and fluid saturation. Geophysics 51, 757−766.

Nedoma, J., 1999. Numerical Modeling in Applied Geodynamics. John Wiley & Sons, New York.

Pearson, C.E., 1974. Handbook of Applied Mathematics. Van Nostrand Reinhold Co, Landon.

Qing-Jiu, Q., 1985. The Operator Theory and Application of Fourier's Integral Operator (in Chinese). Shanghai Sci-Tech Publishing House, Shanghai.

Qing-Yi, C., 1979. Equations of Mathematical Physics (in Chinese). People's Education Pub. House, Beijing.

Riley, K.F., 1974. Mathematical Methods for the Physical Sciences. Cambridge University Press.

Rong-Jun, Q., 2008. The Characteristic of the Seismic Wave and Relevant Technological Analysis (in Chinese). Publishing House of Petroleum Industry, Beijing.

Roseman, D.H., 1985. Nonlinear inversion, statistical mechanics, and residual statistics estimation. Geophysics 50, 2797–2807.

Sabatier, P.C. (Ed.), 1985. Basic Methods of Tomography and Inverse Problems. Adam Hilger.

Schuster, H.G., 1987. Deterministic Chaos. VCH.

Sheriff, R.E., Geldart, L.P., 1983. Exploration Seismology. Cambridge University Press, Cambridge.

Sheriff, R.E., 1984. Encyclopedic Dictionary of Exploration Geophysics. Society of Exploration Geophysics.

Shi-Tong, D., 2009. Theory and Methods of Seismic Wave Mechanics (in Chinese). China Petroleum University Press, Qingdao.

Silvester, P.P., 1983. Finite Element for Electrical Engineers. Cambridge University Press.

Sjöostrand, J., 1980. Propagation of analytic singularities for second-order Dirichlet problems. Commun. Partial Differ. Equ. 5 (2), 187–207.

Sobolev, S.L., 1964. Partial Differential Equations of Mathematical Physics. Pergamon Press, Oxford.

Spencer, A.J.M., 1980. Continuum Mechanics. McGraw-Will, New York.

Tarantola, A., 1984. Inversion of seismic reflection data in the acoustic approximation. Geophysics 49, 1259–1266.

Tarantola, A., 1987. Inverse Problem Theory, Methods of Data Fitting and Model Parameter Estimation. Elsevier Publishing Co.

Telford, W.M., Geldart, L.P., Sheriff, R.E., Keys, D.A., 1990. Applied Geophysics, second ed. Cambridge University Press.

Tikhonov, A.N., Arsenin, V.Y., 1977. Solutions of Ill-posed Problems. Wiley.

Tikhonov, A.N., Samarskii, A.A.S., 1963. Equations of Mathematical Physics. Pergamon Press, Oxford.

Ulrych, T.J., Sacchi, M.D., Woodbery, A., 2001. A Bayes tour of inversion: a tutorial. Geophysics 66, 55–69.

Wen-Cai, Y., Chang-Qing, Y., 2009. The micro-scale reflection wavefield and recognition of hydrocarbon reservoirs (in Chinese). J. Oil Gas Technol. 31 (6), 1–10.

Wen-Cai, Y., Jian-yuan, D., 1992. Seismic scattering tomography with application to crustal investigations. Earthquake Res. China 6 (2), 221–239.

Wen-Cai, Y., Jian-yuan, D., 1993. Approaches to Solve Nonlinear Problems of Seismic Tomography. Acoustic Imaging, vol. 20. Plenum Press, New York, 591–604.

Wen-Cai, Y., Zhen-Min, J., Chang-Qing, Y., 2008a. Seismic response to natural gas anomalies in crystalline rocks. Sci. China Ser. D 51 (12), 1726–1736.

Wen-Cai, Y., Zhi-Qin, X., Chang-Qing, Y., 2008b. Reflection attributes of paragneiss in the upper crust. Sci. China Ser. D 51 (1), 1–10.

Wen-Cai, Y., 1989. Geophysical Inversion and Seismic Tomography (in Chinese). Geo. Pub. House, Beijing.

Wen-Cai, Y., 1990. Seismic velocity imaging of inhomogeneous layered medium. Chin. J. Geophys. 33 (2), 265–282. Allerton Press Inc., New York.

Wen-Cai, Y., 1993. Nonlinear chaotic inversion of seismic traces. Chin. J. Geophys. 36, 241–257.

Wen-Cai, Y., 1997. Theory and Methods of Geophysical Inversion (in Chinese). Geo. Pub. House, Beijing.

Wen-Cai, Y., 2003. Flat mantle reflectors in eastern China: possible evidence of lithospheric thinning. Tectonophysics 369 (3–4), 219–230.

Wen-Gui, L., 1989. Geophysical Inverse Problems (in Chinese). Science Press, Beijing.

White, J.E., Mikhaylova, N.G., Lyakhovitskiy, F.M., 1976. Low frequency seismic waves in fluid-saturated layered rocks. Phys. Solid Earth 11, 654–659.

White, J.E., 1983. Underground Sound Application of Seismic Waves. Elsevier Scie. Pub. B.V.

Wu, R.-S., Maupin, V., 2007. Advances in Wave Propagation in Heterogeneous Earth. Elsevier. A.P, Amsterdam.

Wu, R.-S., Aki, K., 1990. Scattering and Attenuation of Seismic Waves, vols. 1–3. Birkhauser Verlag, Basal.

Xun, D., 1985. Introduction of Continuum Mechanics (in Chinese). Tsinghua University Press, Beijing.

Yan-liang, N., Wen-Cai, Y., Yong-gang, W., 1995. Iterative linearized method for cross-hole tomography. Chin. J. Geophys. 38 (2), 283–292. Allerton Press Inc., New York.

Index

Note: Page numbers followed by "f" denote figures; "t" tables.

Printed and bound by CPI Group (UK) Ltd, Croydon, CR0 4YY

08/05/2025

01864871-0001